THE HAND-BOOK TO BUILDING A CIRCULAR ECONOMY

DAVID CHESHIRE

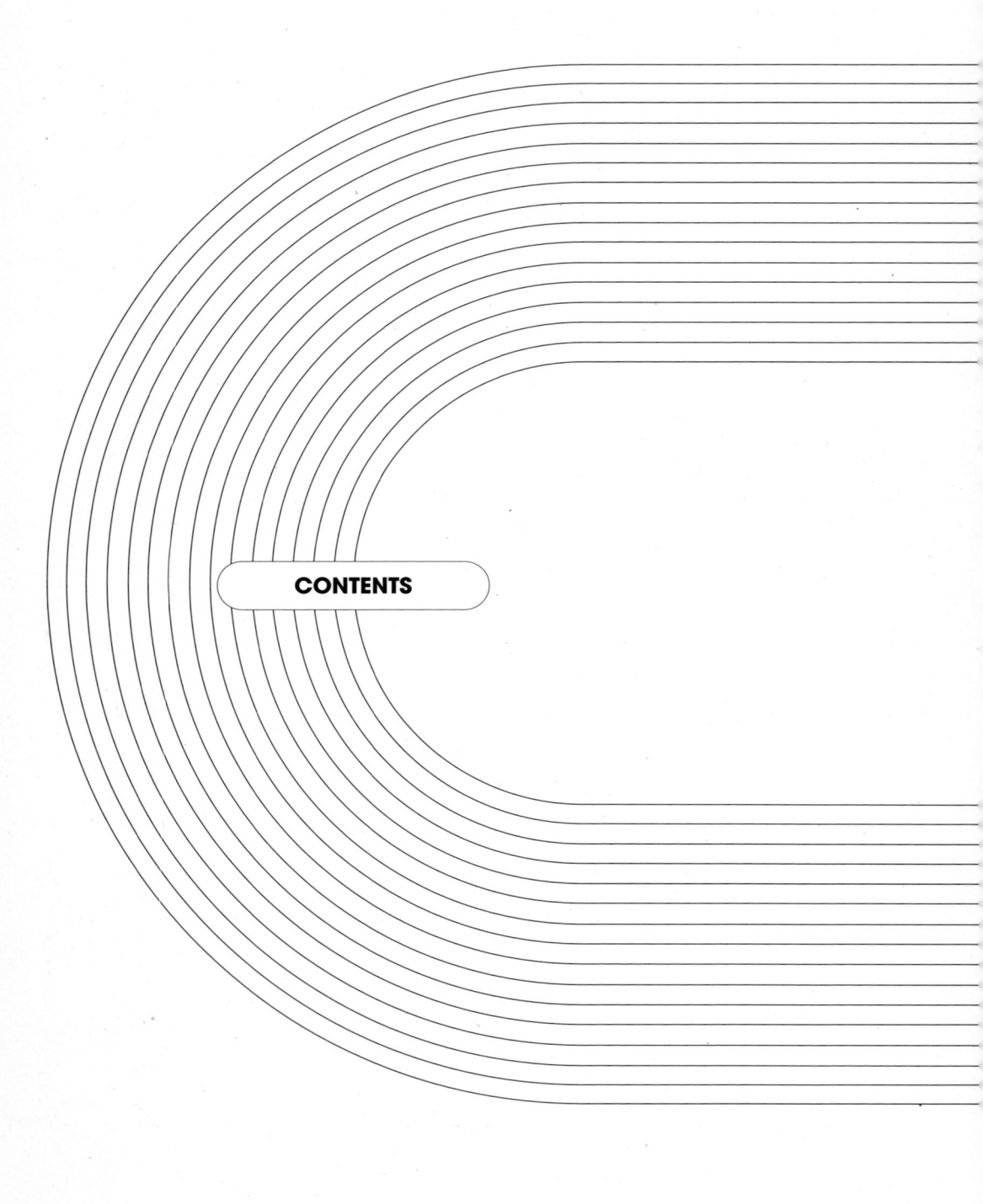

CONTENTS

DEDICATION

The idea for this book was inspired by reading *Cradle to Cradle: Remaking the Way We Make Things* (by William McDonough and Michael Braungart) and the Ellen MacArthur Foundation report, 'Towards a Circular Economy'. The Cradle to Cradle book convinced me that design could be used to make buildings '100% good' rather than just 'less bad' and the Ellen MacArthur Foundation report made me realise that there is an immediate and compelling argument for implementing a more circular economy.

I'd like to thank the contributors who so generously shared their work, case studies, images, ideas and time to help me create this book. I'm also grateful to RIBA Publishing's excellent commissioning and editorial team.

Lastly, I'd like to thank my wife, Jolanta, for her support and for giving me the time to write.

ABOUT THE AUTHOR

David Cheshire is a Director at AECOM, specialising in sustainability in the built environment. David has over twenty-five years' experience acting as a sustainability champion on construction projects and has written best-practice industry guidance, including the CIBSE's *Sustainability Guide* and *Resource Efficiency of Building Services*. David trained as a building surveyor and has an MSc in Energy and the Built Environment. He is a Chartered Environmentalist and an advocate of the transition to a circular economy. The new Spatial Strategy for London[1] includes new circular economy policies for projects based on the principles set out in this book, and David is working with the Greater London Authority to help to implement the new circular economy policies and guidance for major projects in the city.

"A circular economy is an industrial system that is restorative or regenerative by intention and design."

— Ellen MacArthur Foundation

This book explains how buildings can be designed for a more circular economy and how new business models can be used to enable the transition – a crucial step in averting a climate catastrophe. It proposes a set of design principles that can be applied to keep precious resources in circulation by thinking about the whole life of a building, and how the materials and components can be used – or reused – instead of consumed and discarded.

This handbook is an updated second edition of 'Building Revolutions'. This edition includes many new case studies and examples of solutions that can be adopted to help to avert the climate crisis.

IS IT THE END OF THE LINE FOR A LINEAR ECONOMY?
The disposable society is plundering the world of precious, finite resources at an increasing rate.[1] Buildings are stripped out and torn down with astonishing regularity, while new buildings are constructed from components made from hard-won virgin materials. There is little – or no – consideration of the environmental and social impacts of winning and processing construction materials, and there is no real thought about the fate of our buildings once they

have been built. With the global awareness of the climate crisis, the impact of resource use on the climate and the planet's biodiversity is now starting to come into focus.

The built environment demands almost half of the world's extracted materials[2] and waste from demolition and construction represents the largest single waste stream in many countries.[3] The raw materials required for our built environment are becoming harder to extract,[4] and are putting more strain on the environment as fragile ecosystems are exploited.[5] This is combined with geopolitical forces that cause price volatility and disruptions in the supply of essential raw materials. Meanwhile, demand for resources is predicted to rise as the global middle class is set to double in size by 2030.[6]

Global cement consumption is projected to increase by 12–23% by 2050,[7] while steel production is forecast to grow by 30% over the same period.[8]

Building design is, all too often, tailored to the current users and constructed with little thought of the future. Buildings are also composed of complex components with a bewildering number of different materials and polymers melded together irretrievably. This lack of regard for the future life of buildings risks leaving a legacy of obsolete architecture with precious resources locked up and away from future generations.

These systemic problems represent a huge challenge for the construction industry, and the predicted growth in demand for raw materials is going to put further pressure on a resource-constrained world.

THE BIRTH OF A NEW ECONOMIC MODEL

A new model has emerged: one that moves away from the current linear economy – where materials are mined, manufactured, used and thrown away – to a more circular economy where resources are kept in use and their value is retained.

For buildings, this means creating a regenerative built environment that prioritises retention and refurbishment over demolition and rebuilding. It means designing buildings that can be adapted, reconstructed and deconstructed to extend their life, and that allow components and materials to be salvaged for reuse or recycling. Designing buildings for a circular economy can increase their value by avoiding depreciation, and can help to stave off obsolescence. It can even secure a positive residual value at end of life.

New business models allow short-lived elements of the building to be leased instead of purchased, providing occupants with increased flexibility, new relationships with manufacturers, and the ability to procure a service rather than having the burden of ownership. Building collaborative relationships enables suppliers and manufacturers to invest in product development instead of having to focus on the next sale. Creating a demand for reclaimed or remanufactured components will stimulate the local economy and create new industries, while reducing waste.

A forensic examination of the materials that go into buildings ensures that any potentially harmful substances are purged from the design. This improves internal air quality as well as allowing biological materials to be returned safely to the biosphere.

A POSITIVE LEGACY

This book interprets the concepts of a circular economy into a simple set of design principles and business models that can be directly applied to the built environment. The fully updated case studies demonstrate how delightful spaces can be designed from reclaimed products and new materials that are good for the environment and for occupants, as well as saving on construction and materials costs. There are case studies that show how buildings that are designed to be adaptable to different uses (loose fit, long life) can retain value for longer by being more resilient to change in the market. There are also examples of buildings that are designed to be completely demountable, allowing owners to flex their portfolio to changing market needs, alter their function and even move them to new sites in different configurations.

Chapter 13 and its case study (Park 20|20) demonstrates how a developer is reaping the rewards of moving to a more circular economy, by creating a virtuous circle that has reduced construction costs, increased rental yields, sidestepped depreciation and created buildings with residual value. The developer has formed collaborative partnerships with the whole supply chain that has promoted innovation over lowest-cost tendering, and transparency of both cost and the make-up of each building component. Full accountability for the constituent parts of a building has created demonstrably healthier, more valued buildings. Most importantly, it leaves a positive architectural legacy for future generations.

1

WHAT IS A CIRCULAR ECONOMY?

The essence of the circular economy lies in designing goods to facilitate disassembly and re-use, and structuring business models so manufacturers can reap rewards from collecting and refurbishing, remanufacturing, or redistributing products they make.

—Ellen MacArthur Foundation

The principles of the circular economy mean keeping materials and resources in use and retaining their value, rather than consuming and disposing of them. To achieve this, products are designed to have longer lives, and to be reused, remanufactured or reassembled instead of discarded. Alternatively, products have to be made from biological materials without any toxic chemicals so they can be recycled or returned to the biosphere. Waste should be redefined as a resource to be exploited by local industries. As far as possible, new products should aim to design-out waste during their manufacture and at end of life, and to ensure that residual waste can be reclaimed for reprocessing. A transition to a more circular economy means creating new industries and business models that focus on retaining the value of products and materials, and that redefine ownership by providing services rather than selling products. Ultimately, these new reuse and remanufacturing industries should be fuelled by renewable energy.

The current wasteful and inefficient linear economy involves taking raw materials, manufacturing them into products and then throwing them away, often before they have failed or reached the end of their life. Some of these materials may be recycled, but valuable elements such as copper or aluminium are lost forever. It is currently impossible to separate copper or aluminium contaminants from remelted scrap iron or steel.[1] This lowers the quality of the recycled iron or steel and means that the copper and aluminium are lost as a resource.

Some definitions of a circular economy

A circular economy is an industrial system that is restorative or regenerative by intention and design. It replaces the 'end-of-life' concept with restoration, shifts towards the use of renewable energy, eliminates the use of toxic chemicals, which impair reuse, and aims for the elimination of waste through the superior design of materials, products, systems, and, within this, business models.

Ellen MacArthur Foundation, 2013[2]

The UK Government describes a circular economy as: 'moving away from our current linear economy (make–use–dispose) towards one where our products, and the materials they contain, are valued differently; creating a more robust economy in the process.

Environmental Audit Committee, 2014[3]

The European Commission proposes that we move to a more circular economy:

This means re-using, repairing, refurbishing and recycling existing materials and products. What used to be regarded as 'waste' can be turned into a resource. All resources need to be managed more efficiently throughout their life cycle.

European Commission, 2014[4]

This new economic model seeks to ultimately decouple global economic development from finite resource consumption. It enables key policy objectives such as generating economic growth, creating jobs, and reducing environmental impacts, including carbon emissions.

Ellen MacArthur Foundation, 2015[5]

THE ORIGINS OF CIRCULAR THINKING

Walter R. Stahel is a Swiss architect and was an early advocate of the circular economy. He wrote a paper in 1982 called 'Product Life Factor' that proposed 'an economy based on a spiral–loop system that minimizes matter, energy-flow and environmental deterioration without restricting economic growth or social and technical progress'.[6] Figure 1.01 shows the 'self-replenishing system' from that original paper, which encapsulates many of the principles of a circular economy. In the diagram, the inner loop represents the reuse of goods and components (loop 1), the next loop represents repair (loop 2), the third is for remanufacture (loop 3), and the outer loop is for materials recycling (loop 4).

This shows that the concept of the circular economy is by no means new. The ideas are founded on several different schools of thought such as 'cradle to cradle', 'biomimicry' and 'industrial symbiosis'. It takes inspiration from biological cycles where nothing is wasted, and waste is food.

THE HANDBOOK TO BUILDING A CIRCULAR ECONOMY

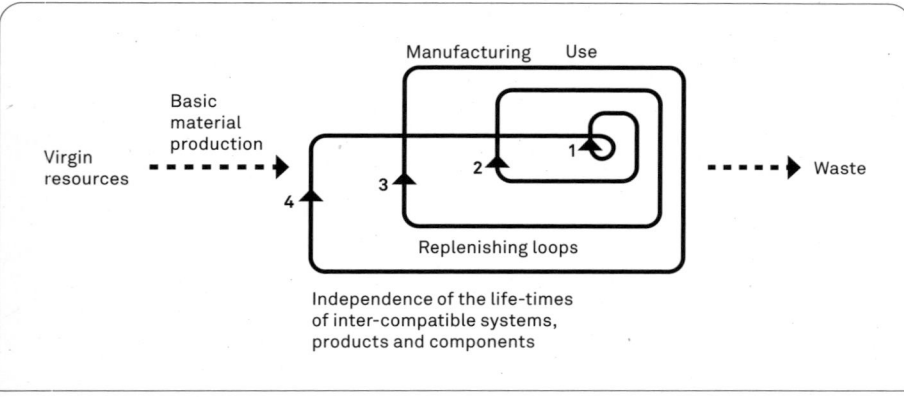

Figure 1.01: The self-replenishing system (product-life extension). In 1982 Walter R. Stahel proposed 'an economy based on a spiral–loop system that minimizes matter, energy-flow and environmental deterioration without restricting economic growth or social and technical progress'.

The ideas behind the current interpretation of the circular economy are closely related to the Cradle to Cradle® design framework developed by William McDonough and Michael Braungart and explained in their seminal book *Cradle to Cradle*.[7] The Cradle to Cradle® philosophy goes beyond circles and closed loops – for the authors, it is about creating a spiral of improvement.

Cradle to Cradle proposes the idea that instead of being 'less bad' to the environment, the aim should be to be '100% good'. It draws on the analogy of the cherry tree that 'makes copious blossoms and fruit without depleting the environment'.[8] It nourishes the soil, provides oxygen, absorbs carbon dioxide and provides habitats for many other organisms. The authors warn against 'eco-efficiency', as this can lead to buildings that have less access to daylight and views in a bid to reduce solar gains, or that are tightly sealed without openable windows to reduce air leakage. Instead, they propose a building that has plenty of daylight and views, openable windows and night-cooling strategies – an 'eco-effective' building.[9] They cite examples of buildings that they have designed that have delighted occupants, including a factory that has a brightly lit, tree-lined internal street with impressive staff retention rates.[10]

Cradle to Cradle suggests that environmental 'regulation is a signal of design failure'.[11] There is no need to regulate to stop a pollutant entering the atmosphere if the need for that pollutant is designed-out from the start, or it can be fed back into the manufacturing process rather than being released. The authors worked with a furniture manufacturer to design a new fabric for upholstery that did not contain synthetic dyes and chemicals. Instead they designed a fabric that was free from toxic chemicals and used a biological material that could be composted at its end of life. This not only created a better product for users, but the costs of disposing of the off-cuts from the factory were avoided, and the effluent from the factory was as clean as the water going in. As an added bonus, the rooms required for hazardous chemical storage could be turned into recreation and work space, and protective equipment for the workers was no longer required.[12]

This idea applies directly to the construction industry: some fit-out contractors are exclusively using water-based paints for all their work, not because they are trying to reduce their environmental impact but because it avoids the need to store hazardous chemicals and protects their staff from the fumes given off by solvent-based paints. This has knock-on benefits: the occupants will not have to breathe in toxic fumes, and the waste will be less contaminated when the fit-out is torn out and replaced.

The authors of *Cradle to Cradle* have developed a certification scheme for products based on the principles they have developed. The scheme labels products based on a detailed assessment of the chemical ingredients in each material and their impacts on health and the environment. See Chapter 10 for more detail.

BIOMIMICRY

Biomimicry is about mimicking the way that living organisms solve design challenges to produce sustainable solutions. The concept of the circular economy and cradle-to-cradle (C2C) philosophy both take their inspiration from living systems, where waste from one process becomes food for another organism, and even poisonous substances can be processed and neutralised.

Inspirational examples of biomimicry in relation to the built environment are described in Michael Pawlyn's excellent book, *Biomimicry in Architecture*, now in its second edition.[13] The book includes examples of how structural efficiency can be improved by mimicking the structure of bones, and even ideas about how buildings can be grown. See Chapter 10 for more on biomimicry and how some of the philosophy can contribute to a circular economy.

INDUSTRIAL SYMBIOSIS

Industrial symbiosis is the study of how materials flow between industries, and finding opportunities to use 'waste' from one industry as a raw material for another.

The National Industrial Symbiosis Program (NISP) has been operating since 2003, and is a world first. Its aim is to inspire businesses to keep resources in productive use for longer. The NISP website (www.nispnetwork.com) includes many examples of industrial symbiosis in action, such as ceramic waste from the pottery industry becoming aggregate for a construction company, resulting in cost savings for both industries, reduced carbon dioxide emissions and waste.

PRINCIPLES OF A CIRCULAR ECONOMY

An interesting interpretation of a circular economy has been made by the Ellen MacArthur Foundation in a series of reports that include an evolved set of principles and compelling arguments for making the transition.[14] The reports include a diagram that captures and summarises the principles, as shown in Figure 1.02. The principles behind this diagram and how they apply to buildings are explained in this chapter and throughout this book.

From the very start, the aim should be to design-out waste and to think about products and materials as a precious resource to be preserved, rather than being wasted after they have been 'consumed'.

The diagram shows two distinct circles, to differentiate between 'biological materials' and 'technical materials'.

Biological materials, such as wood and natural fabrics, are used in products that are free of contaminants and toxins so that they can be returned to the environment at end of life. The diagram shows the potential for 'cascades', so a timber beam could be used in the building structure, then reused as a non-structural element, then formed into a fibreboard before being composted to generate biogas.

Technical materials, such as metals and plastics, are used in products that keep these materials in use for as long as possible, and are retained within the 'technosphere' (technical cycle). This recognises that these materials are essential to the economy, that they are finite and that they should not be returned to the environment, as they may be toxic. Products made from technical materials should be designed to be durable and adaptable, allowing them to be upgraded or re-engineered as necessary. These products should be designed to allow them to be disassembled at end of life. The components should be reused as far as possible and the constituent materials recycled as a last resort.

Buildings contain many complex components, particularly the plant and equipment used to service the building, and the products used in the interior fit-out. As an example, fluorescent lamps contain many technical materials in very small doses, including rare earth phosphors, tungsten, nickel, zinc and aluminium. They also contain toxic substances such as mercury and chromium.[15] In the UK, lamps should be returned to dedicated recycling organisations once they are finished with, but in 2017 lamp recycling rates was only 48.5%.[16] This is a far higher collection and recycling rate than most consumer electronic goods, but it shows how difficult it is to close the loop. Once a lamp is purchased, it becomes the responsibility of the consumer to dispose of it correctly, and this does not always happen. This means that important and scarce metals and toxic substances are being lost or leached to the environment.

One of the ways to keep technical materials circulating is to shift away from the idea that consumers own products and towards the concept that people are users who buy a service. When people purchase a product they are then responsible for its maintenance, repair and disposal. When users buy a service, the manufacturer has a vested interest in the whole life of the product and is ultimately responsible for its disposal. This should mean that manufacturers are incentivised to design products that can be repaired, upgraded and disassembled at end of life. This would allow them to keep the components and the materials in circulation for longer and significantly reduce the amount of material that is wasted. The idea of buying performance rather than products is discussed in Chapter 12.

Reusing buildings and components

Roughly speaking, half of the embodied carbon in a building is tied up in the foundations and the structure.[17] Embodied carbon represents the carbon emissions released from extracting and manufacturing materials and elements, so it does not represent all of the environmental impacts from constructing a building. It ignores issues such as the scarcity of the material and its ability to be

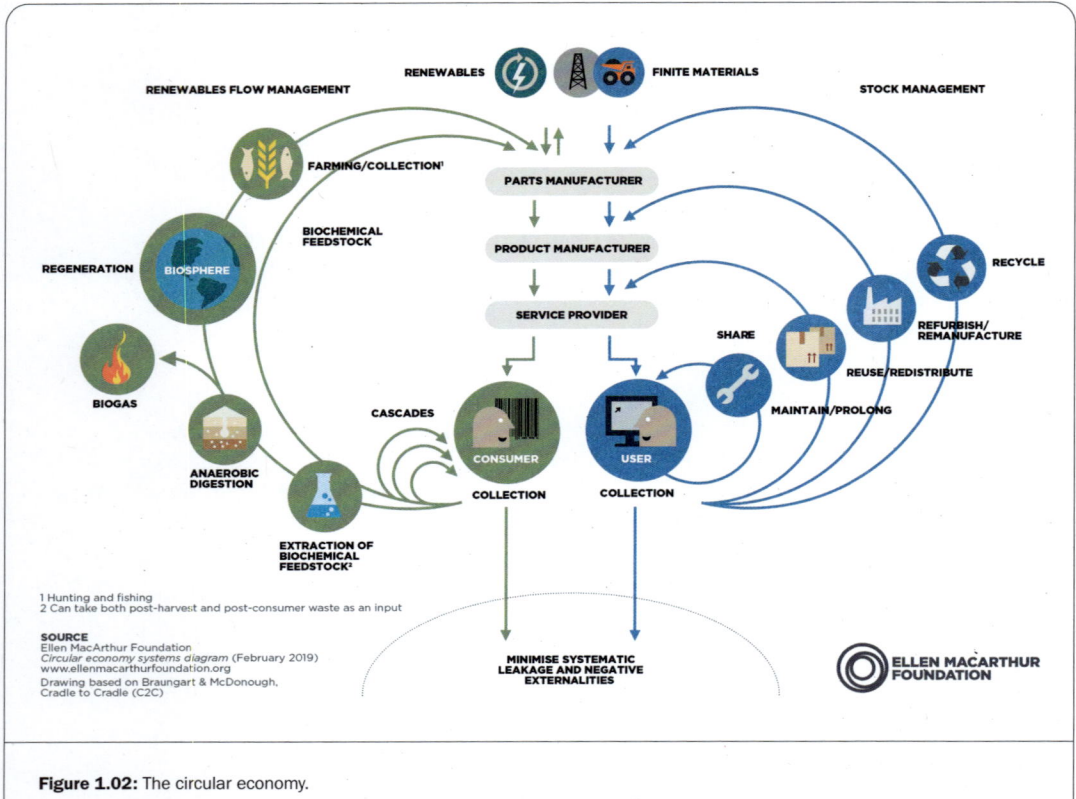

Figure 1.02: The circular economy.

recycled at end of life. However, as a proxy for the resources needed to construct a building, it shows that reusing the foundations and structure of buildings makes a lot of sense.

There are models for designing buildings to be more flexible and adaptable to enable them to have a longer life (see Chapter 8). However, there are lots of reasons why buildings become obsolete, and refurbishment is not always an option.

Once all the options for refurbishing the current building have been considered and the building is going to be demolished, then the next question should be: can the components be reused?

The scrap value of the materials in components is a tiny fraction of the value of the whole product. Reusing rather than reprocessing a component will also have far lower environmental impact. As noted by Bioregional, reusing steel sections can reduce the environmental impacts by 96% when compared to using new steel sections.[18] The difference between the scrap value and the value of the whole product is even higher for complex assemblies such as chillers or boilers.

There is additional cost and time associated with deconstructing buildings, rather than simply demolishing them, but this can be addressed through redesign, and rethinking the process of demolition and rebuilding. This is discussed further in Chapters 9 and 11.

Remanufacturing

Remanufacturing of plant and equipment is more common in other industries, and it is an opportunity that has multiple benefits. According to a World Economic Forum report,[19] Renault has a remanufacturing plant in Choisy-le-Roi, France, that re-engineers different parts including water pumps and engines and then sells them at 50 to 70% of the original price, with a one-year warranty. The operation generates $270 million annually and employs 325 people.

In the construction industry, there is a limited market for refurbished boilers and there are, of course, architectural salvage yards and websites trading in a selection of valuable components – Belfast sinks, chimneypots and the like. In general, though, it is far more likely that building elements are scrapped and new ones are bought for the next project. An example of remanufacturing of furniture is described in Chapter 12.

Recycling

The outermost loop of the diagram is about enabling the recycling of materials back into the manufacturing process. This is the last resort once durable products have been through the other loops to extend their life as far as possible, retaining value and reducing the environmental impacts of reprocessing.

The design of the components should allow the materials to be readily separated for recycling, and they should be clearly identified to allow them to be sorted and reprocessed. The materials should be kept as pure as possible, avoiding contamination from other materials, to allow them to be reprocessed into components of a similar grade.

The current practice is for materials to be downcycled rather than recycled, meaning that they end up as lower-grade products. So concrete becomes secondary aggregates and solid timber becomes particle board.

Current practices and the potential for change are explored in Chapters 4 and 11, respectively.

Avoiding leakage

The bottom of the diagram shows how incineration of waste (even for energy recovery) and landfill is considered 'leakage', as it is wasting valuable resources that leach out of the circle and are lost from the economy. In the UK, the construction industry has been successful in diverting waste from landfill with a

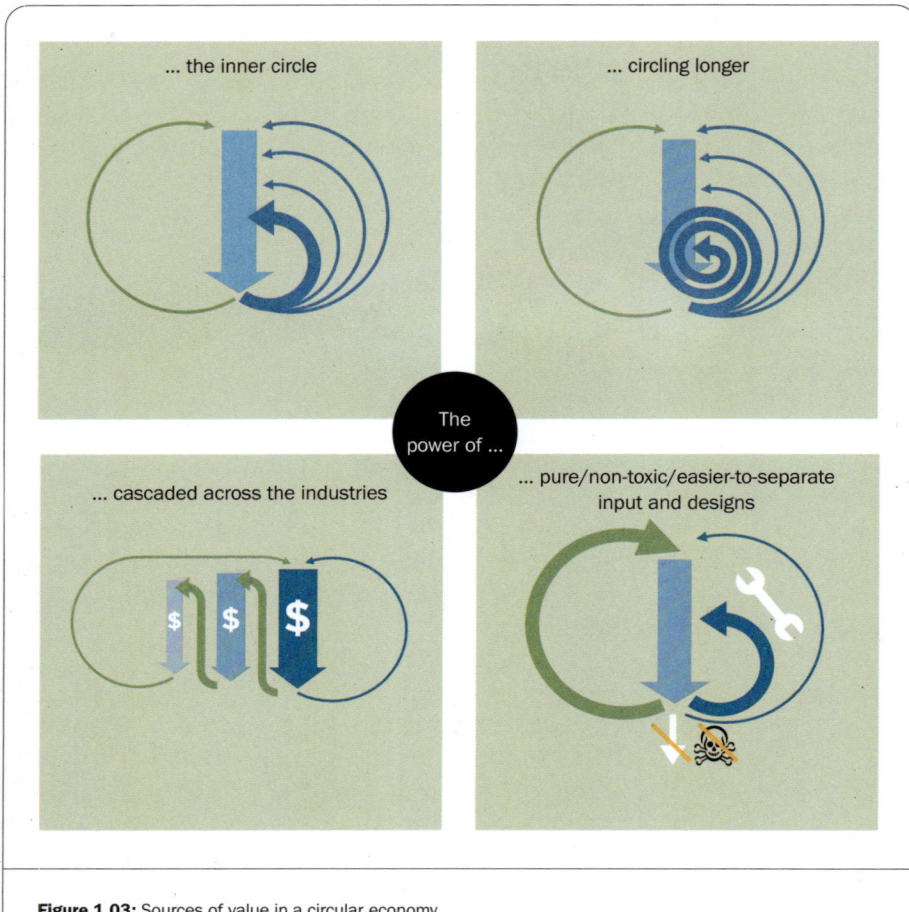

Figure 1.03: Sources of value in a circular economy.

high percentage being recycled in some way, but the residual waste is typically incinerated, meaning that it is lost forever.

Retaining value

The ways that the circular economy model can create and retain value are discussed in the Ellen MacArthur reports and summarised in Figure 1.03. To retain the value of products, the circular economy model requires that:[20]

- the inner circles are prioritised, as they use the least resources. For buildings, this means refitting and refurbishment over demolition and rebuild.

- there are more consecutive cycles, so buildings are refitted, adapted and refurbished instead of being pulled down. And, ideally, when they reach the end of their useful life, they can be disassembled into modules that can then be reassembled in a new configuration and location. This is explored further in Chapters 6 to 9.
- there are cascading uses across industries, with waste products from one industry becoming a useful product for construction, and waste from buildings replacing the use of virgin materials in another industry. Biological materials should be cascaded through uses before being returned safely to the biosphere.
- the materials should be kept pure and uncontaminated to allow them to be reused or recycled into products of equal or better value, or to be composted at end of life.

Applying all of these ideas to buildings will help to reduce the demand for raw materials while saving money, reducing the exposure to the volatile prices of resources and creating new, local industries.

CONCLUSION

The current idea and principles of a circular economy have evolved from many schools of thought into a coherent and highly compelling model that is starting to be applied by innovative organisations and individuals in many different industries. Pioneers are making bold commitments to circular economy design principles on projects, and policymakers are implementing planning policies to drive change. Chapter 5 shows how the principles can be applied to the built environment, while the case studies in this book demonstrate the practical implementation of the ideas in various projects.

A shift to a more circular economy means not only using fewer resources, but also creating and retaining value in buildings and their components. It should not inhibit design, but be used as a catalyst and inspiration for designing buildings that are a positive legacy for future generations. The shift will create new business models and industries that offer ways for organisations to create long-term relationships with customers and provide local employment opportunities from refurbishment and remanufacturing.

WHY CREATE A CIRCULAR ECONOMY?

We have an uncertain future because our economy runs from a manufacturing perspective. We take a finite number of resources, we make something out of them and ultimately, at the end of their life, we throw them away.

—Dame Ellen MacArthur

Awareness of the climate crisis has galvanised the world into action. Many of the efforts to meet the challenge of climate change are focusing on reducing energy demand and the shift to renewable technologies. However, according to the UN's International Resource Panel,[1] the extraction and processing of materials, fuels and food[2] make up approximately 50% of the total greenhouse gas (GHG) emissions. On top of that, resource extraction and processing cause 90% of water stress and biodiversity loss.

It is projected that population growth and per-capita income growth will create a doubling in global primary materials use between today and 2060,[3] with increasing urbanisation driving a demand for infrastructure and buildings.

The built environment demands almost half of the world's extracted materials, and over the next 40 years the world is expected to build 230 billion sq. m in new construction – the equivalent to adding a city the size Paris every week.[4] This growth is projected to drive resource use from 40 billion tonnes to roughly 90 billion tonnes by 2050.[5]

The Ellen MacArthur Foundation[6] estimates that a circular scenario for the built environment could reduce global carbon dioxide emissions from building materials by 38% or 2 billion tonnes in 2050, due to a reduced demand for steel, aluminium, cement and plastic.

Population growth and the escalating demand for consumer products are driving up the demand for resources. It is predicted that, by 2030, three billion people who are currently living in poverty will join the middle-class level of consumption, which would create a corresponding surge in the demand for resources. According to the Ellen MacArthur Foundation, even the more conservative projections suggest that demand for natural resources will increase by at least one-third.[7]

This increasing demand is creating price volatility and disruptions in the supply of essential raw materials, caused by conflicts, disasters, pandemics or by countries restricting the international trade in raw materials, such as when China restricted the export of rare earth elements. This restriction created a sharp hike in prices of products that relied on these elements. Notably, in 2011, a price increase of over 300% was reported for rare earth phosphors used in fluorescent lamps.[8] With China still controlling over 90% of the rare earth market, the rest of the world is still vulnerable to supply disruption and insecurity in the market.[9]

Environmental issues are the underlying reasons for many of the supply disruptions, and climate change is causing more extreme weather events, adding to the problems.

Winning and processing resources is putting increasing pressure on the environment as these essential raw materials get harder to extract. Mining and drilling in more remote locations requires more energy, water and materials to access, and poorly controlled activities can result in the release of toxins that destroy farmland, fragile ecosystems and people's health.[10] Even pristine ecosystems like the Amazon rainforest are mined for precious materials such as bauxite (for making aluminium). These mining activities lead directly to deforestation and the destruction of habitats and have other impacts, including an increase in soil erosion, pollution and the fragmentation of habitat caused by building access roads.

To compound the problem, it is becoming increasingly resource-intensive to extract fossil fuels, with offshore wells now having to be more than twice as deep as they were in the late 1990s.[11] This increases the demand for steel and means that more energy is required to extract a diminishing amount of fuel.

As noted by Julian Allwood and Jonathan Cullen in *Sustainable Materials: With Both Eyes Open*, most ores require the extraction of ten tonnes of rock to gain just one tonne of ore, and then the element has to be extracted from the ore. This extraction requires energy and uses chemicals, some of which are harmful.[12] Accidental or incidental emissions of these chemicals to soil, water and air can have serious and long-lasting impacts on the environment, surrounding species and local people. The extraction of raw materials has also been linked directly to conflicts in many parts of the world.

Copper is a good example of a key metal that is heavily used in buildings, including cladding, wiring, electronics, plumbing, heating and cooling pipework, heat exchangers in refrigeration, and so on. Most copper mining is located in arid areas, which puts increasing strain on scarce supplies. It is estimated that there was about 7% of copper in every tonne of ore 150 years ago; now it is more like 0.6%.[13]

An analysis of the water used to produce important materials identified those materials produced in areas of high water scarcity, which require high levels of water to produce.[14] These include common building materials such as iron, copper and aluminium, which are all mined in areas of water scarcity. It is expected that water constraints will limit production in Australia, South Africa, China, India and Chile. On top of this, climate change is expected to reduce water availability in all these countries. It is predicted that water availability will drive price increases for raw materials in the future.

The supply disruptions have prompted many countries to identify the materials that they consider 'critical' to their industries. A European Commission report on critical raw materials identifies twenty materials that are considered critical because of both their high relative economic importance and their high relative supply risk.[15] According to the EU report, the reason that the raw materials have a high supply risk is because a large share of the worldwide production comes from only a few countries.

The construction industry accounts for over half of the UK's resource use, by weight,[16] and is dependent on materials such as aggregate, cement, clay products, timber, lead, copper, glass, iron and steel for its structure and fabric.

The insightful Absolute Zero report[17] maps out a route to achieving genuinely zero emissions (i.e. not one that uses offsets such as planting trees) by 2050. It notes that, in construction, the most impactful materials are steel and cement. Therefore to achieve absolute zero, all steel used in construction would have to be from recycling, and there would have to be considerably less use of cement.

Buildings are now bristling with technology, with electronic components integrated into many products. This technology depends on the use of many critical materials, in small but essential amounts. Even the permanent magnets in motors for lifts, chillers, heat pumps and the like use rare earth elements such as neodymium iron boride to make them as energy-efficient as possible.[18]

BENEFITS OF A MORE CIRCULAR ECONOMY

There are significant benefits in moving to a more circular economy. In the UK, Defra's waste statistics showed that 63% (120 million tonnes) of the waste stream was attributed to construction, demolition and excavation waste in 2016.[19] Even a small increase in reuse and remanufacture would save tonnes of resources and millions of pounds. The Chartered Institution of Waste Management estimates that reducing waste across Europe could save €72 billion per year and create over 400,000 new jobs in Europe.[20]

These are just the benefits of reducing waste in the system. Another key benefit of a more circular economy is to increase repair and remanufacturing of products; estimates suggest there is potential for this industry to grow by more than £3 billion, with a corresponding increase in skills and employment opportunities.[21]

Businesses and industries that move to a more circular economy should reap important benefits if they embrace the concepts fully:

- Careful management of materials and resources will help to save money through closer links with the supply chain and a better understanding of resource risks.
- Companies can maintain control and even ownership of materials, helping them to protect themselves against volatile prices and provide more security of supply.
- There should be significant reductions in the cost of compliance with environmental legislation and the disposal of waste.
- Most radically, the move away from selling products towards providing a service will keep companies more in touch with their customers (and building occupants!) for longer, providing more opportunities to sell-in more services and to better understand the needs of the customer.

These new business models can result in more repair, reuse and recycling of products. Rolls-Royce has been applying a service-based model for many years by offering 'power by the hour', which covers 'full in-use monitoring, servicing, repair, remanufacture and replacement' of its engines.[22]

These new business models can have secondary benefits as well. When Philips started selling a lighting service, called Pay-per-lux, the leasing arrangement included incentives for Philips to closely manage the lighting controls to keep the energy consumption as low as possible.

IT'S ALREADY HAPPENING

The Ellen MacArthur Foundation and McKinsey management consultants identify three trends that are contributing to a transition to a more circular economy:

- First, resource scarcity and tighter environmental standards are here to stay. Their effect will be to reward circular businesses that extract value from wasted resources over take–make–dispose businesses.
- Second, information technology is now so advanced that it can trace materials anywhere in the supply chain, identify products and material fractions, and track product status during use.
- Third, we are in the midst of a pervasive shift in consumer behaviour: a new generation of consumers seems prepared to favour access over ownership.[23]

These trends are driving companies to change their business models and take advantage of the emerging markets and potential financial benefits. Some companies in the construction industry, notably carpet manufacturers including Desso, Interface and Shaw, are already implementing circular economy thinking and reaping the rewards, including designing carpets that can be readily recycled and creating leasing models that allow customers to buy a flooring service rather than owning a product. Desso, a major Dutch carpet manufacturer, has embraced cradle-to-cradle principles (see Chapter 1). According to Desso's CEO, the strategy has delivered increased profit margins despite the global economic crisis, indicating that customers are already willing to pay a premium for the greener product lines.[24]

Caterpillar has had a remanufacturing business since 1973. It manufactures and sells machinery and engines, including construction and mining equipment. It handled more than 70,000 tonnes of remanufactured products in 2010 and has been growing at a rate of 8–10%, according to the Ellen MacArthur Foundation[25] (see Chapter 12).

Cities as far apart as Brussels, Phoenix, Toronto, Singapore, London, Copenhagen and Amsterdam are all working towards a transition to a circular economy. The 2021 London Plan[26] (the region's statutory spatial development strategy) includes new circular economy policies for projects, based on the principles set out in this book (see Chapter 5). The London Plan requires a Circular Economy Statement to be prepared for all major planning applications, which demonstrates how the proposed development prioritises retention and refurbishment over new buildings and responds to the circular design principles. It requires the preparation of a Whole Life Carbon Assessment that covers embodied carbon and operational emissions.

CONCLUSION

In the looming shadow of the climate crisis, there is a pressing need to reduce the demand for raw materials and the waste arising from the construction and demolition industries. The circular economy model brings financial benefits for companies that adopt it, as well as reducing the environmental and social cost of a consumer society.

The shift to a more circular economy is happening, with manufacturers seeing the benefits and offering new products and services, including some within the construction industry.

It is easier to see how a manufacturer will gain from applying circular economy principles, as it can benefit financially from maintaining more control over its supply of raw materials and reducing both waste and the cost of complying with environmental legislation. Consumer products are also relatively short-lived, so designing them for remanufacture or reprocessing will pay off in a few years, and new financial models can help to secure longer-lasting relationships with customers.

But buildings are not consumer products, and they are not 'manufactured' in the same way as other products. This makes it more difficult to see how the circular economy model can be applied to buildings and the construction industry, with its complex and fragmented supply chain, long lifetimes and complete disconnect between construction and demolition. The next chapters explain the current situation and then show how circular economy principles really can be applied to the built environment.

BUILT TO LAST?

We build to endure, to resist time, although we know that ultimately time will win. What previous generations erected for eternity, we demolish.

— Habraken

When a new building is being designed and constructed, it is hard to imagine that it will one day be demolished. It is harder still to expend the time and effort to think about whether the building can be designed so that the materials and components can be reclaimed for reuse or recycling at the end of life.

Commercial buildings are often refurbished or demolished before their structure and fabric fails. The reasons for demolition are more likely to be related to changing land values, lack of suitability of the building for current needs or lack of maintenance of various non-structural components.[1]

Buildings such as laboratories become obsolete quickly due to rapid changes in technology or new research techniques, whereas high-quality housing can remain current and useful for longer.

Buildings that we value do not depreciate. Quite the reverse: they often have increasing value attached to them as they become older. The stock of listed buildings shows how the buildings that are really valued and loved are retained over their contemporaries.

On the other hand, Habraken points out that 'conservation may serve to freeze works of art in time, resisting time's effects. But the living environment can persist only through change and adaptation'.[2] Conservation can be criticised for stifling the natural complexity of the built environment and inhibiting its natural adaptation. This raises the question of whether the built environment should be continually refreshed and adapted, or retained and preserved as far as possible.

A paper by Forest Lee Flager contrasts Notre-Dame cathedral in Paris with the Ise shrine in Japan.[3] Gothic cathedrals across Europe are visited by millions of tourists a year to marvel at the feats of architecture and engineering, and the longevity of the buildings. Flager argues that the 'evolution of the Gothic cathedral is conveyed, at least in part, through the physical materials used in its construction. The years of effort required to bring the vision of Notre-Dame to completion are evident in the craftsmanship of every detail'.

This contrasts with the ancient shrine complex in Ise, Japan. Known as the Ise Jingu, the shrines were originally built in the ninth century and have been rebuilt over 60 times since (see Figure 3.01) to reflect the principles of Shinto and wabi-sabi, which hold that impermanence and ageing are intrinsic to all natural materials. The inner shrine, Mishine-no-Mikura, at Naiku, Ise, and the outer shrine, Taka-no-Miya, at Geku, Ise, are dismantled every twenty years and reconstructed on adjacent sites using the same design, with new materials, as part of a ceremonial ritual that lasts seventeen years (see Figure 3.02). The buildings are reconstructed using traditional techniques before the originals are dismantled and their constituent parts distributed to other holy shrines and grounds for reuse.

The building is constructed from Japanese cypress, with a raised floor, a roof thatched with miscanthus grass, and supporting timber pillars buried in the

Figure 3.01: The Mishine-no-Mikura shrine at Naiku, Ise (adjacent to the inner sanctuary), in Japan.

ground. In an article, Junko Edahiro states that the rebuilding ceremony is an important national event and explains:

> *Its underlying concept – that repeated rebuilding renders sanctuaries eternal – is unique in the world. In the occidental way of thinking, creating something durable would normally involve building a structure with robust stones, bricks, and concrete. At this shrine, however, the structures are made exclusively from wood and, by being rebuilt over and over again, can last forever. Also, in the process of rebuilding, the skills of shrine builders and craftsmen in various fields (carpentry, sacred treasures, apparel, etc.) are passed on from generation to generation.*[4]

The shrines are built using elaborate joinery techniques with joints that are flexible but with enough strength to last without the use of nails, screws or adhesives. The connections make allowances for the shrine to be repaired and disassembled at the end of the twenty-year cycle.[5]

Flager quotes the architect Kisho Kurokawa to explain the different philosophies:

Figure 3.02: Taka-no-Miya shrine at Geku, Ise, in Japan, showing the old and new shrines sitting side by side.

We have in Japan an aesthetic of death, whereas you [Westerners] have an aesthetic of eternity. The Ise shrines are rebuilt every 20 years in the same form, or spirit; whereas you try to preserve the actual Greek temple, the original material, as if it could last for eternity.

If buildings are designed to be permanent, then they have to be of a very high quality and be greatly valued by future generations to survive. The alternative is to think of a building as a fleeting, temporary structure that will ultimately be destroyed to make way for the new. This means buildings can become disposable consumer goods, which is faintly absurd when they demand so much resource to build and generate so much low-value waste during refurbishment, refit and demolition.

BUILDING OBSOLESCENCE

The value of a building is a complex interaction between factors such as:

- market forces
- regulations
- technology changes

- build quality
- physical deterioration.

The reasons for a building becoming obsolete and then being demolished may have nothing to do with its physical condition; it may just be because its original function has become redundant. Warehouses in many of the docklands in the UK became obsolete when goods were brought in by containers rather than being offloaded by hand from small ships. Many of these warehouses were ideal for adaptation into other uses, ranging from offices and housing to markets and museums. For example, the original Georgian warehouses in West India Docks in London now house the Museum of London Docklands, restaurants, shops and apartments.

Research shows that the quality and characteristics of a building are a better indicator of deprecation than its age.[6] Baum, et al. notes that building lives are getting shorter, but that 'depreciation is not forever: depreciation for older property is lower than depreciation on new property. Better underlying value is available in older buildings'.[7] The problem comes when the value of the building depreciates to the point that it is close to the value of the land; at that point, it is likely to be demolished and the land redeveloped.

So build quality can be an important way to retain value, as poor build quality will increase the cost of adaptation,[8] making it more likely that buildings will be demolished.

CONCLUSION

Designing buildings to be more flexible and adaptable should allow them to retain their value for longer by avoiding functional obsolescence. Alternatively, buildings can be consciously designed for a short life, and with the ability to be disassembled so that modules can be redeployed in new buildings or so that the materials can be separated and reclaimed for further use.

Either way, designing buildings that are easier to refit, refurbish or dismantle helps to reduce the demand for raw materials and the amount of waste arising. The best way to design for disassembly and reuse is to understand what happens at the end of life, as explained in the next chapter.

STARTING AT THE END

The construction sector has a high raw material dependence and handles materials with high intrinsic value, while generating significant volumes of waste.

— Ellen MacArthur Foundation

Buildings are a complex assembly of materials that are designed for construction and use, with little or no consideration of how they will be adapted and refurbished or demolished at end of life.

Consequently, any changes to the building during its life or at its demolition create tonnes of waste that are typically downcycled into lower-grade products rather than being reused or recycled. That is aside from the fact that buildings are often demolished before the building elements have failed.

Great strides have been made in increasing the amount of construction waste that is diverted from landfill – often over 90% in some countries. But this raises the question: where does the waste go now that it is not being buried in the ground?

CURRENT DEMOLITION PRACTICES

Demolition practices have, in general, moved away from the salvaging and reclamation of materials towards more mechanised demolition and reprocessing of materials. The demolition process has become increasingly mechanised due to the requirement to rapidly clear the site, to improve health and safety standards and to reduce the cost of demolition by using less human labour and more machines. Deconstruction is still undertaken on constrained sites where heavy machinery could affect adjacent structures or cause nuisance to neighbours, and is also used in refurbishment projects if more time and budget is available. Deconstruction can take twice as long and manual methods require more labour and more access on site, which poses further safety risks.

Reclaimed materials have to be sorted, cleaned and indexed, all of which takes time and effort. They then have to be carefully stored and a buyer has to be found. On the other hand, the scrap value of materials is high and the returns are immediate, which helps with cash flow during the demolition contract.[1] It is clear from this why many demolition contractors choose recycling over reclamation.

The 'soft strip' of a building is usually the first activity on site after the initial planning stages and the disconnection of services. The soft strip includes the identification and removal of any hazardous wastes (e.g. asbestos and fluorescent tubes), followed by a first pass to remove fixtures and fittings and then a second pass to remove redundant services. Non-structural elements, such as partition walls, may also be removed at this stage.[2] Metals in the building services and internals, including aluminium, steel and copper, are removed as part of the soft strip and sold for reprocessing. Even complex and valuable building services equipment, such as; chillers, pumps and boilers, is often sold as scrap rather than considered for reconditioning and reuse.

Once the soft strip is completed, the building fabric is then demolished as quickly and efficiently as possible. Steel beams are cut into manageable lengths using cutting torches or shears, and reinforcing bars are segregated

Figure 4.01: Scrapped raised floor tiles. Could these have been reclaimed?

Figure 4.02: Demolition site, showing how a building is torn down, leaving a pile of mixed waste.

from concrete. Composite facade panels are peeled apart to reclaim the metals and complex components. Structural insulated panels with foam insulation containing greenhouse gases (e.g. CFCs and HCFCs) must be sent to specialist processing plants to capture the environmentally damaging gases.

The materials arising from the demolition are either segregated and reprocessed on site or sent to materials recovery facilities (MRFs), depending on the constraints of the site and the processes involved. Concrete, blockwork and hardcore materials can be segregated and crushed on site for use as aggregate. High-quality bricks laid in sand lime mortar may be reclaimed, but the majority are crushed. Timber and timber products, plastics, plasterboard, insulation, ceramics and packaging are typically sent to MRFs for segregation and ultimately for reprocessing.

WHAT HAPPENS IN AN MRF?

In the UK, MRFs process the majority of mixed waste from construction sites. These processing facilities are cavernous industrial buildings containing huge piles of mixed waste and queues of trucks coming to replenish the heaps as they are sorted (see Figure 4.03).

Figure 4.03: Construction and demolition waste at an MRF prior to sorting.

Excavators are used to sift out the bulky waste, including mattresses, large metals and composite products such as electrical equipment. Inert waste, such as aggregate, may be diverted to different facilities for processing. Some waste may be shredded to make it easier for magnets and other automatic machinery to extract materials. The waste is then dropped on to an elevated conveyor belt. The metals are picked out with magnets, and eddy current separators are used for the non-ferrous metals. Soil and other residue are filtered out using a vibratory screen. The next generation of MRFs are becoming increasingly mechanised, with technologies such as optical sorting. Many MRFs still rely on human labour, with the waste being passed through a picking area where a gang of staff (known as 'pickers') grab and throw materials into the relevant bays below the conveyor. The pickers pull out wood, particle board, plastics, any remaining metals and cables as the waste travels along the conveyor. Some facilities also sort hard plastics into broad polymer groups.

After the picking line, another vibratory screen or trommel shakes out any remaining inert waste such as aggregate and concrete. At the end of the conveyor, a new heap of residual waste is formed, representing around 20–40% of the material that was delivered to the site.

Any untreated solid timber is downcycled into panel products or wood fuel pellets. The composite and treated timber is chipped and used as refuse-derived fuel (RDF). Soft plastics are recycled back into other plastic products. The metals are shredded and smelted into (often) inferior-grade metals. The hardcore is crushed and screened to be reused as secondary aggregate.

Some waste materials cause problems in MRFs. Composite products are difficult to process and will often end up as residual waste. Plasterboard that is correctly separated at construction sites can be sent for reprocessing, and can be turned into new plasterboard if it is salvaged as part of a take-back scheme. Otherwise, it can end up as treatment for farmland or compost. Any plasterboard that is broken up and mixed with other waste has to be treated separately and sent to dedicated mono cells at landfill sites. This is because plasterboard gives off hydrogen sulphide if it is disposed of in conventional landfills.

The majority of the residual waste becomes RDF by being shredded, baled in plastic and shipped out to incineration plants in continental Europe, where it provides heat for buildings.

So, a 90% diversion from landfill really means that around 80% of the diverted materials (at best) are recycled, or more accurately downcycled to a lower-grade material, and the rest is incinerated.

FINAL DESTINATION OF WASTE

Site waste management plans typically cover the journey from the site to the MRF and may say 'reuse offsite' or 'landfill capping'. The destination of waste beyond the MRF is rarely tracked by developers, designers and contractors, but

4 Starting at the end

this is not the end of the story. According to the *Waste Duty of Care Code of Practice*[3] 'You have a responsibility to take all reasonable steps to ensure that when you transfer waste to another waste holder that the waste is managed correctly throughout its complete journey to disposal or recovery.' So, although great care and attention are given to obtaining Waste Transfer Notes and ensuring that the waste contractors are properly licensed, whoever creates waste has a responsibility right through to its final resting place, not just for the journey. If waste is illegally dumped or mismanaged, then it is the producers' responsibility.

The rubble, topsoil and clay from digging foundations and basements have to find a new home. For infrastructure projects, this is well managed: the 5 million tonnes of spoil from the Crossrail tunnel in London was used to create a 670-hectare wetland on Wallasea Island in the Thames Estuary. It is far more difficult to handle all the waste from the multiple buildings being demolished and the basements being dug in cities throughout the world.

CONCLUSION

The demolition industry is moving towards increased mechanisation and away from deconstruction to save time and money and to ensure the safety of workers. Selling high-value materials for reprocessing provides a faster return than reclaiming them for reuse. This means that building materials from demolition activities are unlikely to be salvaged for reuse without a radical rethink. The demolition and waste processing industries are simply playing the hand they are dealt. To change, they need better cards. And that means going back to the beginning and rethinking the design of buildings.

The current building design and construction practices are creating a legacy for future generations that will make it increasingly difficult for them to reclaim valuable, uncontaminated materials in an increasingly resource-constrained world.

To bequeath the next generation with buildings that can become 'resource banks' in the future means that designers and constructors have to reverse the current trends in the industry and embrace the principles set out in this book.

CIRCULAR ECONOMY PRINCIPLES FOR BUILDINGS

Look at a building, there are hundreds of tonnes of valuable materials in those buildings. The residual value of those buildings and materials is a negative one – we have demolition cost and the valuation of commercial real estate is based on discounted cash flows. No-one takes (into account) the materials because (buildings) are not designed to be material banks.

— Coert Zachariasse

Applying circular economy principles to the construction sector and buildings offers huge potential rewards.

Buildings can be designed to have a positive, enduring legacy by making them more adaptable and ensuring that valuable materials and components can be reclaimed and reused at end of life. Ensuring that buildings can be disassembled provides the opportunity for them to be redeployed in new places or for new uses, and allows components to be salvaged and reused or remanufactured. This in turn reduces dependence on raw materials for construction, while salvaging and remanufacturing creates local employment. Declaring and understanding the ingredients that make up materials and components will help to ensure that biological materials can be safely returned to the biosphere and technical materials can be reclaimed for reuse within industry. There is also the added benefit that the use of pure materials with the contaminants designed-out helps to provide better environments in which people can live and work.

Figure 5.01 summarises the principles of a circular economy when applied to buildings. This diagram is included in the London Plan (the spatial strategy for London).[1]

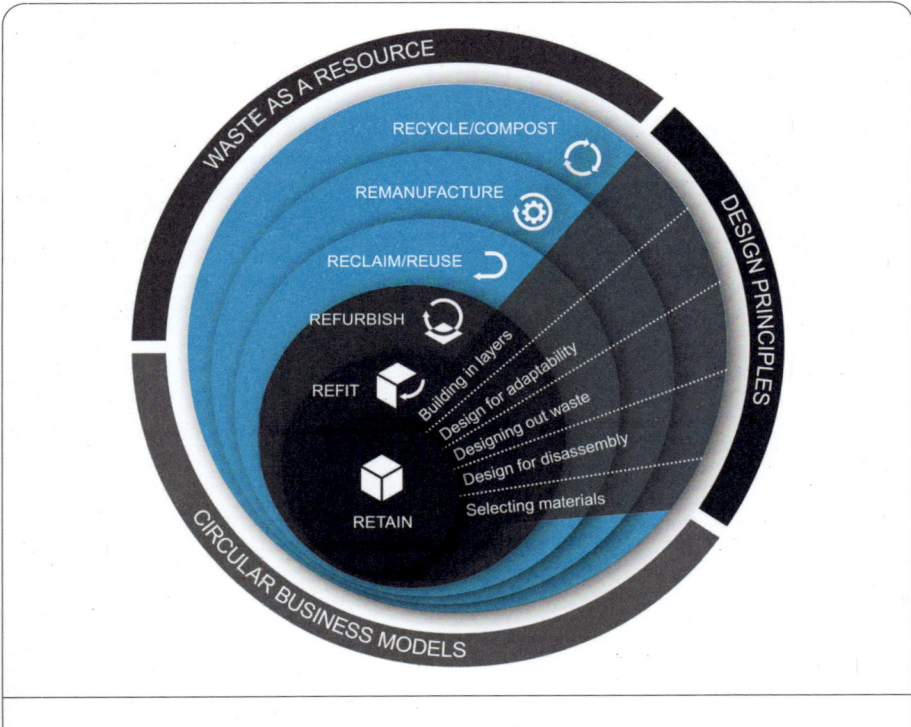

Figure 5.01: Applying circular economy principles to building design.

The nested circles show the hierarchy, with the three inner circles being the most desirable. Retaining the existing building is the most resource-efficient option, followed by refits and refurbishment of existing buildings, as this retains the most resource-intensive parts of the building. For the three outer circles, the priority is to reclaim or remanufacture components, with the last option being to disassemble them to recycle back into new products or return the materials to the biosphere. This hierarchy underpins the design principles covered in this book.

The five segments overlaid on the circles show the design principles that can be applied to reduce waste, extend the life of the building and enable the reclamation of materials at end of life. These design principles are covered in Chapters 6 to 10.

The outer ring in the diagram represents the underlying models that can be applied to enable a more circular economy across the buildings sector, and are described in detail in Chapters 11 and 12.

REFURBISHMENT AS A PRIORITY

Circular economy design principles make a significant contribution to the net zero carbon agenda by radically reducing the carbon emissions associated with construction. The top-down carbon budgets that have been calculated include stringent targets for embodied carbon. For example, the London Energy Transformation Initiative (LETI), a network of over 200 built environment professionals,[2] and RIBA in its 2030 Challenge[3] have proposed embodied carbon targets of less than 600 $kgCO_{2e}/m^2$, which is a tough call given that a typical commercial building would be between 1,000 and 1,500 $kgCO_{2e}/m^2$.

The balance between embodied carbon and operational carbon emissions has shifted in many countries that have a high proportion of renewable energy. In the UK, the decarbonisation of the electricity grid means that all-electric buildings have considerably lower carbon emissions in operation than even ten years ago, and this will fall further as more renewable energy generation capacity is built.

Figure 5.02 is an illustrative comparison of the embodied and operational impacts of a new building and a refurbished building over a 35-year period. The figures are based on current good practice estimates, and the refurbishment scenario assumes that the substructure and structure are retained (representing roughly half of the embodied carbon of the building). The operational emissions gradually decline in line with the UK projected carbon emissions for electricity. The operational emissions for the refurbishment are assumed to be only slightly worse than those of a new building, as the fabric and services are completely replaced.

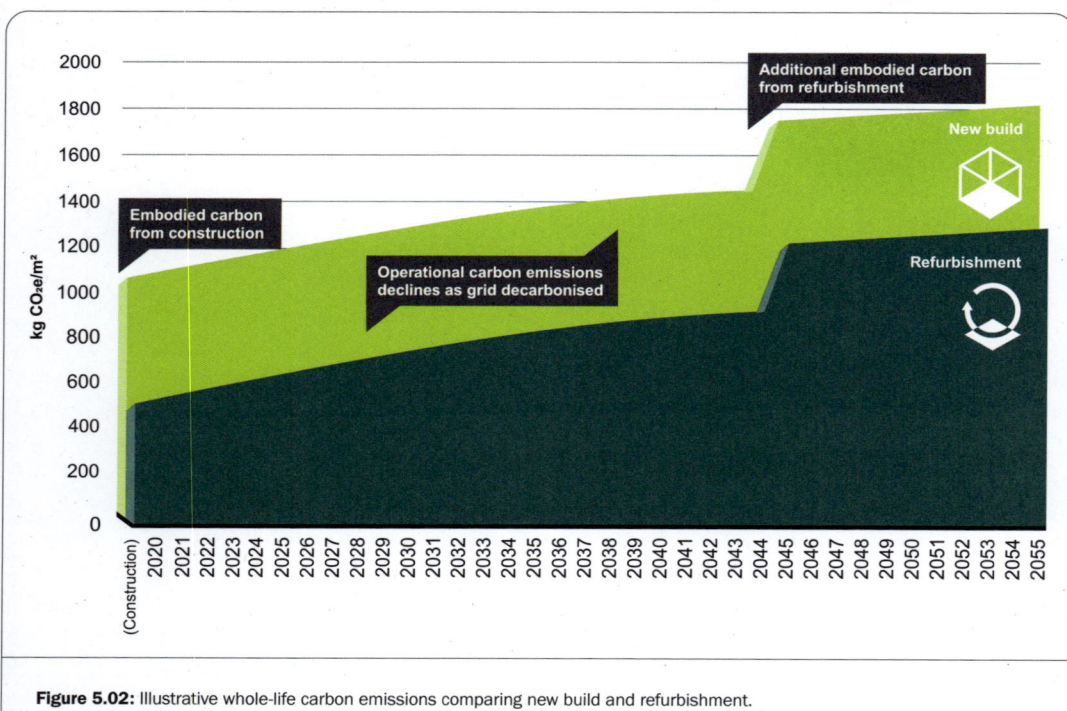

Figure 5.02: Illustrative whole-life carbon emissions comparing new build and refurbishment.

The graph shows clearly that the bulk of emissions for new buildings are front-loaded and twice as high as those from refurbished buildings. The argument that new buildings will pay back their initial, embodied carbon over time by being far more efficient in operation when compared to a refurbished building is no longer valid. Making the best of our existing building stock will help to reduce the risks associated with climate change.

APPLYING THE PRINCIPLES

Table 5.01 shows how the design principles are applied at different points, from the organisational level down to the component choices.

Implementing the design principles in isolation on one project is difficult and time-consuming. It is far more effective to establish the approach at an organisational and portfolio level. Organisations can then establish partnerships with members of the supply chain who offer circular business models. Organisations with a building portfolio can consider reclamation and reuse across many buildings, and the use of consolidation centres to store and distribute items and materials.

The organisation can build the business case for applying the principles outside of the budget and time constraints of a live project. In fact, applying circular economy principles can reduce capital costs through alternative business models. For example, partnering with the supply chain can reduce construction time as demonstrated in the Park 20|20 case study in Chapter 13. This is also helped by eliminating wet trades, which turns construction into onsite assembly. Pay-per-use models move capital expenditure to operating costs, as shown in the Mitsubishi Lifts case study in Chapter 12. Lean design principles go hand in hand with capital cost savings, and resource-efficient, net zero carbon buildings should require less equipment to be installed. Reclaimed and remanufactured components can be cheaper than new, and just as good, as demonstrated by the remanufactured raised floor tiles discussed in Chapter 11.

Applying circular design principles at every level, from policy to practice, has the potential to save time, capital costs and operational costs, and retain the value of the building for longer.

REGENERATE CIRCULAR ECONOMY DESIGN TOOL
Based on the design principles set out in this book, researchers at the University of Sheffield, in collaboration with the author, have developed a new circular economy design tool called *Regenerate*. The tool uses a series of circularity criteria split into four categories:

- design for adaptability
- design for deconstruction
- circular materials
- resource efficiency.

These criteria are then applied to the core building layers: site, structure, skin, services and space, which can be used for all building types such as retrofits and new builds. The tool takes the user through a series of questions that prompt design teams to consider different design approaches, scores the current performance and highlights potential enhancements. Figure 5.03 shows how the tool is structured. The tool was piloted during 2020 and has been launched as an online version in 2021.[4]

CONCLUSION
The overarching philosophy is to put thought into the destiny of the building and how it will serve the next generation.

Buildings are often considered as a lasting legacy for future generations, but all too often they become a liability that has to be demolished at a cost. Commercial buildings often depreciate over time until they have no residual value. But buildings that are valued by people do not depreciate, and can often

Point of intervention		Circular Economy principles		
		Desiging out waste	Design for disassembly	Design for adaptability
	Organisation approach	Partner with offsite / modular construction firms	Strategy for disassembly and reclamation	Decide level of adaptability for all buildings
	Portfolio strategy	Prioritise refurbishment / refit of existing buildings	Re-deployment of buildings	Establish potential to adapt portfolio of buildings
	Design strategy	Design out waste (e.g. lightweight design, reclaimed elements)	Building designed for disassembly	Decide level of adapability (e.g. use scenario modelling)
	Systems selection	Select resource efficient systems (e.g. modular components)	Select / design systems for disassembly	Select / design adaptable systems (e.g. relocatable partitions)
	Component choices	Select recycled / reused components	Select components that are designed for disassembly	Design adaptable components (e.g. magnetic signage)

Table 5.01: Applying circular economy design principles at different points of intervention.

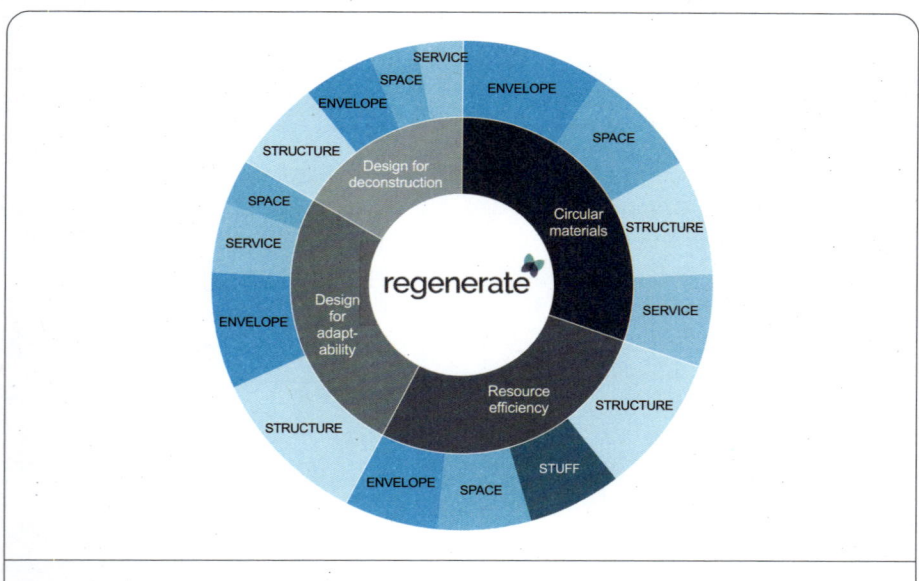

Figure 5.03: The Regenerate tool, illustrating the structure of the Regenerate tool.

Materials selection	Waste as a resource	Business models
Partnerships with alternative materials suppliers	Materials passports and exchange platforms	Partnerships with supply chains to encourage new business models
Reclamation and reuse across portfolio	Provide consolidation centres and map local resources	Consider sharing / leasing of space and services
Design based on alternative materials choices	Use reclaimed elements from other industries	Explore alternative ownership models
Alternative systems (e.g. fabric ductwork)	Select remanufactured or reclaimed systems	Consider leasing systems or using manufacturers with incentivised return
Alternative components (e.g. Cradle to Cradle Certified®)	Select materials made from waste from other industries	Consider leasing components or extended producer responsibility

appreciate over time. By thinking about the potential future life of a building and learning lessons from the buildings that have endured, designers can create buildings that are more flexible and adaptable, giving them a longer life. Alternatively, designers can deliberately design for a short lifetime and ensure that elements of the building can be readily disassembled and reused at end of life. There is even the potential to design buildings that can be demounted and reassembled in new locations and reconfigured for new uses. This is particularly relevant to organisations that experience rapid change in customer demand or numbers of people who need to be accommodated.

By thinking more about the whole life of the building, designers can create a lasting positive legacy for future generations, gifting them either adaptable buildings or giving them readily accessible materials and components with which to create new buildings.

6

BUILDING IN LAYERS

Our basic argument is that there isn't any such thing as a building. A building properly conceived is several layers of longevity of built components.

— Frank Duffy

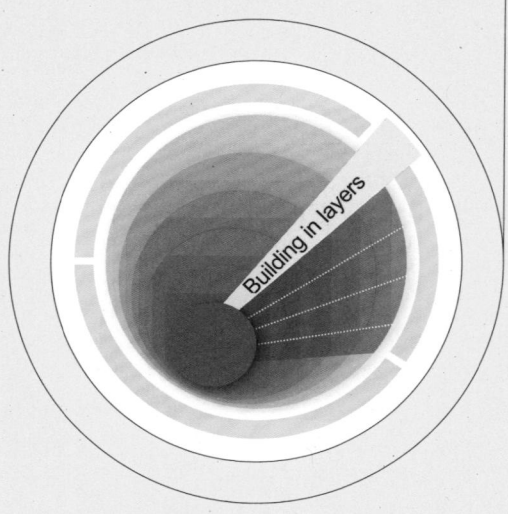

The lifetime of different elements of a building can vary from well over a hundred years down to a matter of months, or even weeks. The structure and fabric of a building, when made from durable materials such as stone, slate, brick or concrete, has been proven to last for hundreds, if not thousands, of years. Newly installed components can fail, prematurely, in a few weeks or be torn out when they do not meet the new occupants' needs or expectations. This calls for a different approach to designing structure and fabric as compared to those elements with a shorter life.

There has to be a clear delineation between the elements with different lifetimes, so sealing electrical wiring into walls or encasing them behind durable finishes means that there will be disruption and waste when they need replacing or moving. Equally, some products may have built-in obsolescence, with inaccessible components that fail prematurely.

Frank Duffy proposed a layered approach to building, and identified four layers:

1. shell
2. services
3. scenery
4. set.

The shell is the structure; the services are pipes, ducts and wires; the scenery is the internal fit-out; and the set is furniture, fittings and equipment. Each of these has a different lifespan, with the shell lasting the longest (the life of the building) and the set lasting only a few months, depending on the occupants.

Stewart Brand expanded and adapted this model in his famous book, *How Buildings Learn*,[1] into six layers:

1. site
2. structure
3. skin
4. services
5. space plan
6. stuff.

(See *Figure 6.01*).

In Brand's categories, the structure and the skin are separated, which accords with the Open Building philosophy (see Chapter 8). The site is the geographical setting; the structure is the load-bearing elements, such as the foundation and skeleton; the skin is the exterior surface, such as the facade; the services are the circulatory and nervous systems of a building, such as its heating plant, wiring and plumbing; the space plan includes walls, flooring and ceilings; and stuff includes lamps, furniture, appliances and ICT.

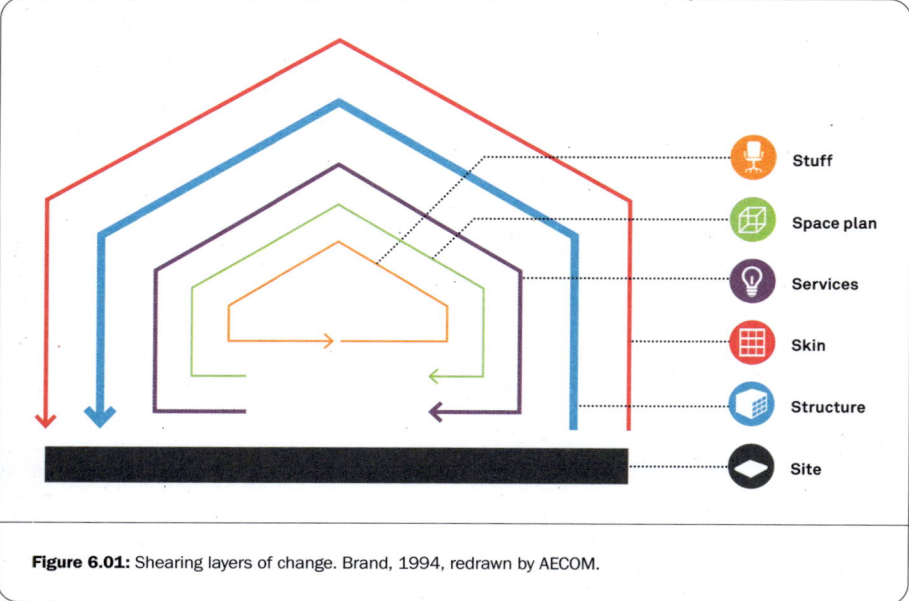

Figure 6.01: Shearing layers of change. Brand, 1994, redrawn by AECOM.

This layering approach is invaluable when considering how to approach the different elements of a building.

Considering the lifespans of different layers and components in the building has benefits at each stage in a building life. During construction, it can help with the sequencing of tasks, and broadly follows the construction process (e.g. structure is followed by skin, then the internal fit-out). During the operation of the building, careful separation of components dependent on actual lifespan should help to make the short-lived components more accessible for replacement. Provision should be made to access components that require regular maintenance, repair and upgrade.

During refits or refurbishment, the ability to peel off layers and apply new ones ensures that the neighbouring layers are undamaged. In one way, the layering approach needs some refinement: it is very likely that building services will need to be maintained or replaced before the finishes (space plan) are changed, and finishes such as tiling could have a far longer lifetime than the services buried behind them.

Creating buildings that are more flexible and adaptable to other uses also draws on the idea of layering the building by ensuring that the primary structure is independent from the secondary structure, allowing for more interventions.

Finally, the use of independent layers helps to enable disassembly at the end of life of the building by allowing each element to be removed independently.

Figure 6.02 shows how the different layers can be separated based on the intended life of each element.

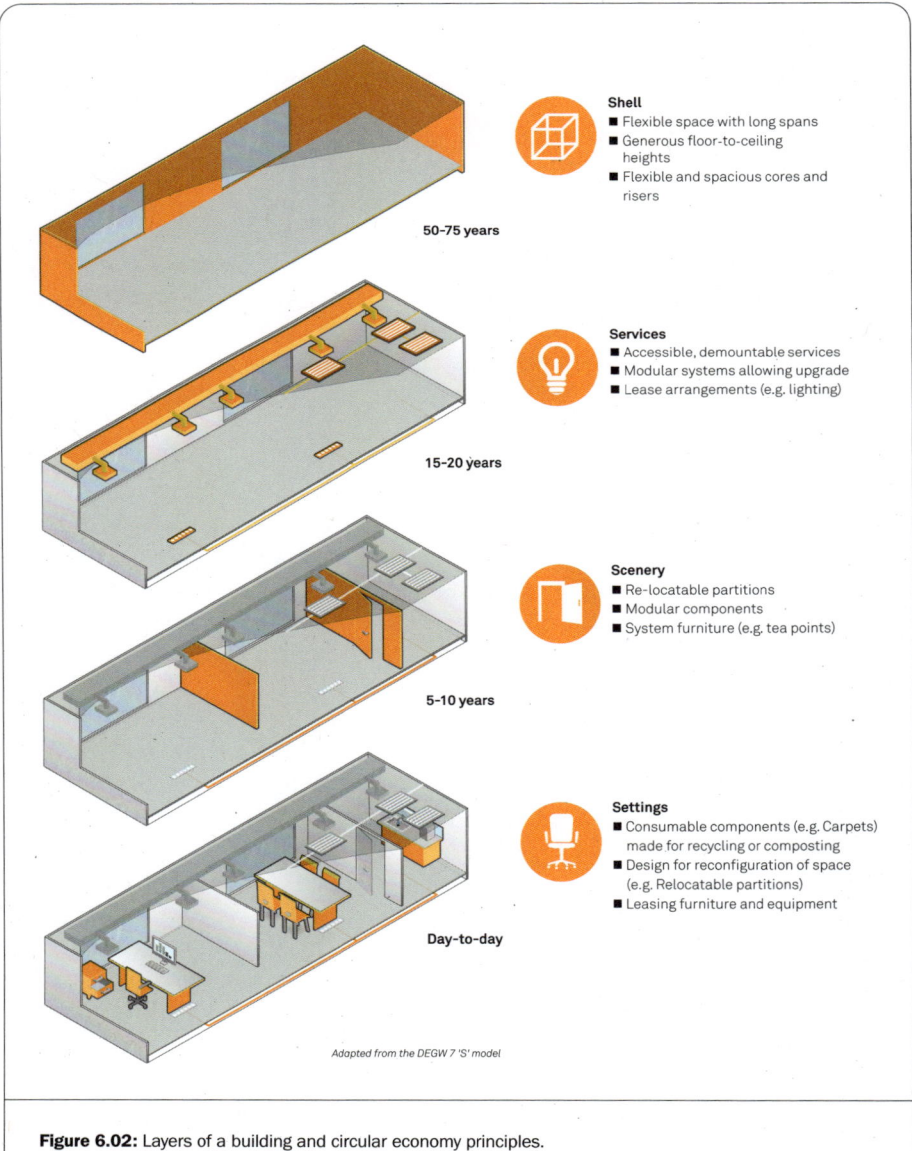

Shell
- Flexible space with long spans
- Generous floor-to-ceiling heights
- Flexible and spacious cores and risers

50-75 years

Services
- Accessible, demountable services
- Modular systems allowing upgrade
- Lease arrangements (e.g. lighting)

15-20 years

Scenery
- Re-locatable partitions
- Modular components
- System furniture (e.g. tea points)

5-10 years

Settings
- Consumable components (e.g. Carpets) made for recycling or composting
- Design for reconfiguration of space (e.g. Relocatable partitions)
- Leasing furniture and equipment

Day-to-day

Adapted from the DEGW 7 'S' model

Figure 6.02: Layers of a building and circular economy principles.

The shell

The structure of the building has the longest potential lifespan and is the limiting factor in adapting a building to a new use. If the facades and the internal configurations are independent of the structure, this should enable adaptation. Chapter 8 discusses the principles of designing for adaptability. This includes designing-in some level of redundancy in the structure and providing generous floor-to-ceiling heights, as well as carefully positioned service cores and the capability for new service runs to be installed between floors or across the floorplate.

Ideally, the structure and the fabric need to be independent to allow parts of the facade to be replaced, or a whole new facade installed. The ability to exchange, say, a glazed unit for an inlet louvre or a fully glazed facade with a solid element would make a building more adaptable to different uses.

Within the fabric layer, there are further sub-layers to consider, especially when designing for disassembly (see Chapter 9). This includes ensuring that the different layers of the fabric of the building can be easily separated at end of life. In particular, elements that use composite constructions with integral insulation have proved hard to separate for reuse, or even recycling, at end of life.

Services

Commercial buildings often make provision for accessing services through access panels, suspended ceilings and raised flooring systems. However, services are still often buried in walls, tangled up with the structure or trapped in the building with no provision for removal and replacement. In traditional brick and stone buildings the structure and the facade are combined, and the services are often woven through all of the elements.

Using the layered approach also helps to make the building easier to maintain, as the services will be more accessible for repair and maintenance. The design strategy should identify the elements that need maintenance or upgrade, and ensure that they are accessible and have components that are durable and can be repaired.

Scenery

Scenery – such as partition walls, floor finishes, built-in cabinets and so on – will have a shorter life than the other elements of the building. Refits or reconfigurations of internal spaces generate a considerable volume of waste that is typically downcycled. Scenery should be designed to be reconfigured, reused, relocated or recycled (rather than downcycled), or it can be made from biodegradable materials that allow it to be composted at end of life.

The consumable components such as carpets and furniture can be designed with a shorter life, and made of biological materials that can easily be recycled or broken down at end of life. Several carpet manufacturers already have ranges of carpets that meet these criteria, and furniture manufacturers are designing closed-loop products. For example, chair manufacturers are designing their products so that they can be readily recycled, and it is now even possible to obtain cardboard furniture. Some manufacturers are offering leasing arrangements that mean customers can purchase a service (e.g. lighting or seating) rather than a product. This gives customers more flexibility and allows the manufacturer to retain ownership of the products (see Chapter 12 for more information and some examples).

PEELING THE LAYERS

In *Sustainable Materials with Both Eyes Open*,[2] Allwood and Cullen propose an onion skin model, with the core being the structural frame with a long expected life and the outer layers having progressively shorter lifespans. The idea is that components with shorter lifespans are easily separated while the core can be retained and adaptable enough to allow for future changes. Figure 6.03 shows a simplified version of an onion skin model for a building.

This is a useful way to envisage and analyse the different layers of the building, but it is overly simplistic because elements rarely fall into one lifespan. Figure 6.04 shows how different elements such as roofs, facades and services are made up of both short-lived and long-life components with interfaces between elements that have to be carefully managed to allow disassembly.

To complicate things further, many elements have a mixture of durable and perishable components. When the least durable components fail, the danger is that the whole system is removed and replaced, even if these items are a minor part of the overall materials value.

Facades and glazing systems are a classic example. High-performance double and triple glazing is now an inherent part of buildings, but it typically uses adhesive sealants, which have a far shorter life than the metals, stone and glass used in facades. By considering how the least durable components can be upgraded and replaced it is possible to double the life of the durable materials and retain their value. Figure 6.05 shows the granular layers of a façade system with the range of service lives of the components.[3]

There are ways to solve this problem. 1 Triton Square was built in the 1990s. Twenty years later, British Land decided to adapt it to meet customers' evolving needs. Rather than demolishing it and building new, British Land took a circular economy approach by retaining and reusing as much of the existing building as possible. Most of the existing fabric and superstructure was retained, while

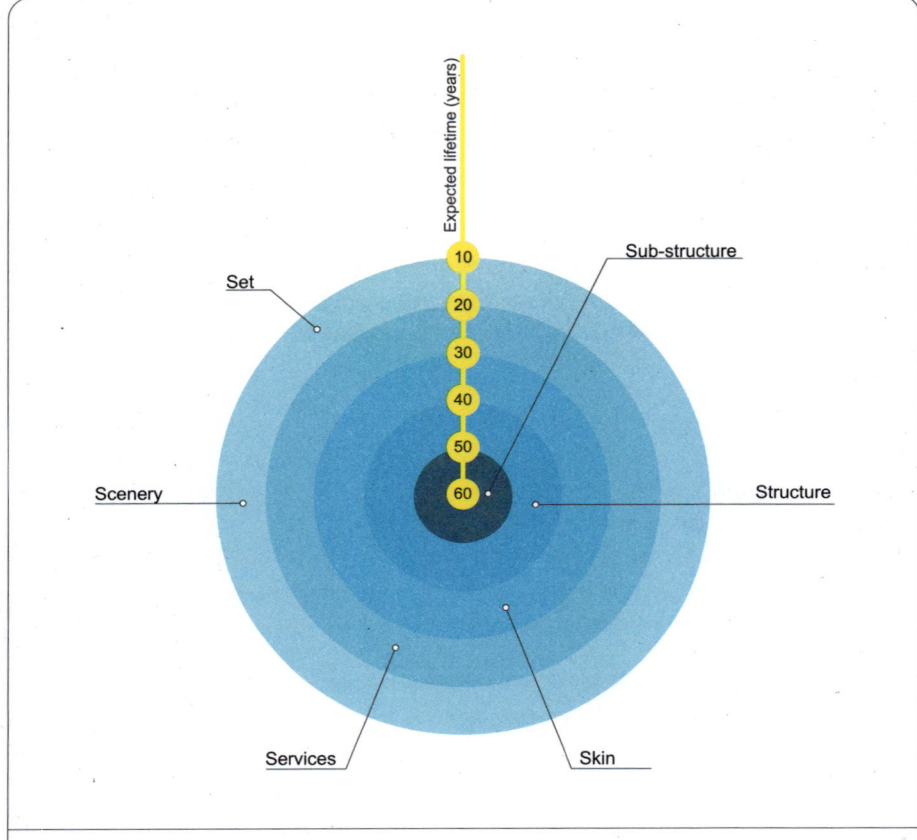

Figure 6.03: Onion skin model for a building with expected lifespans (inspired by Allwood and Cullen's onion skin model).

adding three extra floors and doubling the lettable area of the offices. Instead of scrapping the old facade and replacing it, the design team committed to remanufacturing the old glazing panels. They created a pop-up factory 30 miles away, where they cleaned, refurbished and fitted new gaskets to 3,500 m^2 of glazing panels that were then refitted into the building (see Figure 6.06).

This example shows how designers can reassess the approach to existing buildings to remanufacture components and illustrates the importance of considering sub-layers when designing new buildings.

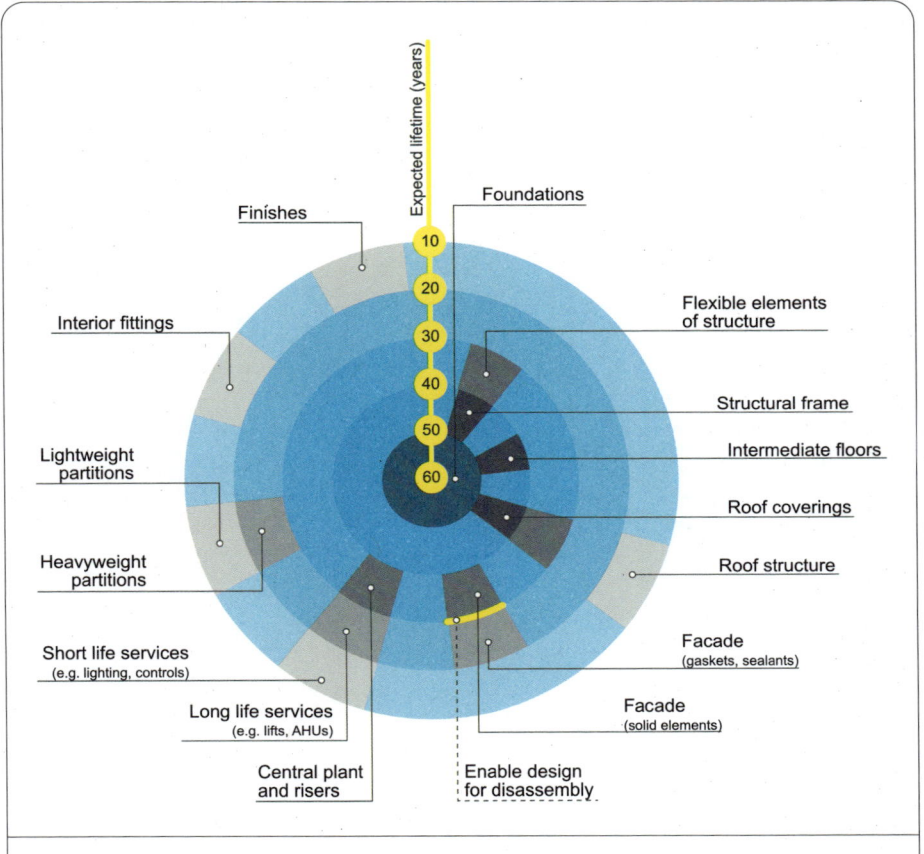

Figure 6.04: Onion skin model with different elements mapped on to the layers with an interface that needs to be managed to enable disassembly.

CONCLUSION

The principles of the circular economy cannot be applied wholesale to buildings, as they are a complex collection of products, all with different lifespans, purposes and demands.

The idea of designing-in layers helps to unlock this tricky problem by allowing designers to approach each element of the building using different rules. So, the structure and fabric can be made to be adaptable and over-engineered to last a lifetime, while the internals will be fickle and have to be designed to be

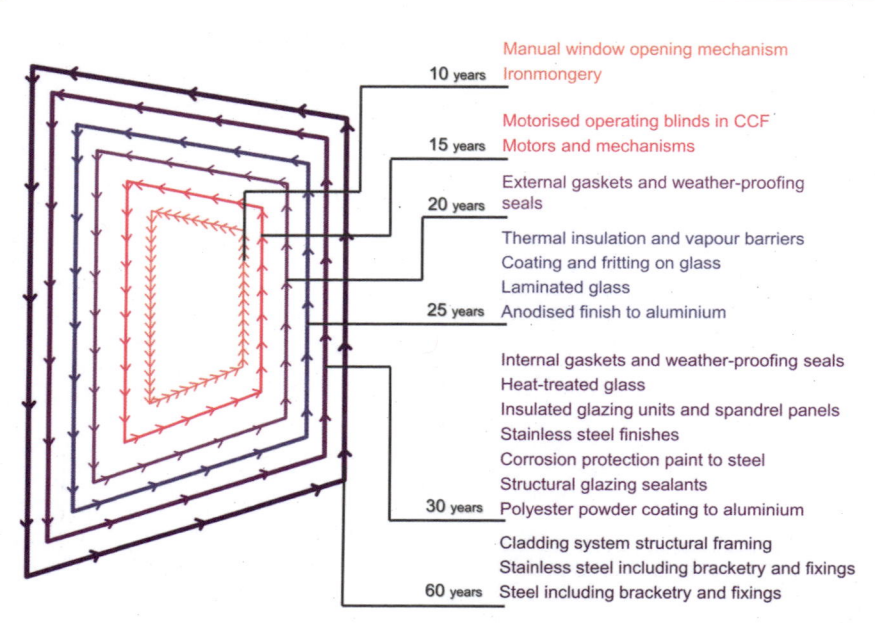

10 years   Manual window opening mechanism
Ironmongery

15 years   Motorised operating blinds in CCF
Motors and mechanisms

20 years   External gaskets and weather-proofing
seals

Thermal insulation and vapour barriers
Coating and fritting on glass
Laminated glass
25 years   Anodised finish to aluminium

Internal gaskets and weather-proofing seals
Heat-treated glass
Insulated glazing units and spandrel panels
Stainless steel finishes
Corrosion protection paint to steel
Structural glazing sealants
30 years   Polyester powder coating to aluminium

Cladding system structural framing
Stainless steel including bracketry and fixings
60 years   Steel including bracketry and fixings

Figure 6.05 Typical facade component service life in layers obtained from industry recommendations (by permission of Rebecca Hartwell), building on Stewart Brand and formerly Frank Duffy's interpretation of the building's 'shearing layers'.

Figure 6.06: 1 Triton Square – remanufactured glazing panels.

reusable or compostable. Similarly, the idea of leasing a structural beam may be far-fetched, but leasing furniture instead of purchasing it makes complete sense.

Identifying the components that have a shorter life than their bedfellows is critical, as there is a danger that the whole element will be scrapped because of one perishable item. Finding ways to either design-out the more perishable items or to allow them to be replaced without damaging the rest of the element is the key.

Keeping each of the layers independent allows the structure to be retained when upgrading the fabric, and the building will be easier to disassemble at end of life so that the components can be reused, remanufactured or recycled.

DESIGNING OUT WASTE

Waste does not exist when the biological and technical components (or 'nutrients') of a product are designed by intention to fit within a biological or technical materials cycle, designed for disassembly and refurbishment.

— Ellen MacArthur Foundation

A circular economy is not just about reducing waste, it is fundamentally designing-out the concept of waste. This means that designers have to think about the whole life of the building, from the decision to build new or refurbish through to the eventual demolition or deconstruction of an obsolete building.

Therefore, the idea of designing-out waste means avoiding creating waste in the first place, and looking for opportunities to turn waste from other places into a resource. For buildings, this includes:

- refitting and refurbishing existing buildings rather than building new
- designing-out waste arising during construction
- using reclaimed materials and components in design
- applying lean design principles to reduce demand for resources and associated waste.

This chapter covers each of these ideas, along with case studies that show how different elements of these principles can be applied.

REFIT AND REFURBISHMENT

The greenest building is the one already built.

Attributed to architect Carl Elefante

When deciding whether to refurbish an existing building or to rebuild, the constraints of the existing building are often cited as the reason for demolition. Typical issues are the floor-to-ceiling heights, riser space or the floor layout. There is also often a financial incentive to build new, as this can increase yields and, in the UK, the tax regime offers little incentive to developers to refurbish existing buildings.[1] On the other hand, there are many situations where buildings are refurbished instead of being demolished, and the designers work around these constraints. In the UK, some developers have built their business models around refurbishing existing buildings into attractive places to live and work.

The Tea Building and the White Collar Factory

For Derwent London, breathing new life into old buildings is a central tenet of its business model. It aims to reuse as much of the fabric of the original building as possible to reduce resource use and time on site, and to save money. The Tea Building in Shoreditch, London, is a great example. Built in the 1930s as a bacon factory for Allied Food's Lipton Tea Brand, it was used as a tea-packing warehouse for most of its life. Derwent London converted it into a series of individual office units by keeping the structure and facade, refurbishing the windows and installing new main plant and services (see Figures 7.01 and 7.02).

Figure 7.01: The Tea Building in Shoreditch, London.

Much of Derwent London's portfolio has been carefully selected to have generous floor-to-ceiling heights, good daylighting and exposed thermal mass to help regulate temperature.

The Tea Building is the inspiration for Derwent London's latest design concept, dubbed the 'White Collar Factory'. According to this concept, Derwent London would avoid wasting resources on fitting out the building in case incoming tenants wanted something different. Instead, it would provide a minimal, efficient initial servicing strategy, giving tenants a high level of flexibility about how they want their spaces to be fitted out and serviced. So meeting rooms and cellular offices could be added as pods to be plugged into the concrete core cooling system embedded in the ceiling slabs and acoustic partitions, along with any other elements, as required. To assist prospective tenants, Derwent London has arranged and costed a range of options to help the decision-making process. Figure 7.03 shows one of the fit-out options.

These examples show that the refurbishment of an existing building can produce results that are as attractive as a new building, and potentially more appealing to occupiers.

55 Baker Street

Sometimes it takes a visionary design team to propose a solution when the only option appears to be rebuilding. In the redevelopment of 55 Baker Street, it was assumed that the huge 1950s office building would have to be demolished to

Figure 7.02: The Tea Building in Shoreditch, London.

Figure 7.03: White Collar Factory fit-out visualisation with mezzanine.

allow the site to be transformed into a new urban centre. Instead, Expedition Engineering (the structural engineers) and Make Architects proposed a solution that would preserve and enhance the existing building as far as possible. The developer, London and Regional Properties Limited, immediately grasped the benefits of the refurbishment option over the new-build option and appointed the team to implement the design.

The floor layout of the existing building was constrained by the eight full-height vertical circulation cores that inhibited the creation of the open-plan floorplates typically demanded by commercial tenants. The structural engineers proposed retaining the majority of the existing reinforced concrete frame and the targeted demolition of the cores to allow the creation of the open floorplates. The cores were replaced with new concrete infill structures to provide lateral stability.

New building structures were inserted into the original 'H block' layout to create a design that includes three atria and infill blocks creating additional floor space and connecting adjacent wings of the original building. The glazed facades provide a new, coherent face to the building along the front elevation (see Figure 7.04).

A steel transfer structure was installed to allow the removal of twelve existing reinforced concrete columns supporting seven floors above, opening up the entrance and reception area (see Figure 7.05).

This project shows how new life can be breathed into a building that is destined for demolition. Around 70% of the original building structure was reused, and the upgrade took a year less than a rebuild solution.[2]

Figure 7.04: 55 Baker Street: before and after refurbishment.

Figure 7.05: Steel transfer structure in entrance.

DESIGNING-OUT WASTE ON SITE

The amount of waste generated on construction sites can be reduced considerably by changing the way that buildings are designed and constructed. The most effective way to reduce waste arising on site is to move more of the construction activities off site and make the work on site more about assembling components than the cutting and shaping of materials.

Even when using traditional construction techniques, there are opportunities to design-out waste. Examples include using modular components (e.g. doorsets rather than doors) and coordination of the structural grid with the external cladding and internal finishes and partitions. The use of 3D design software can be invaluable in ensuring that designs are coordinated and in reducing wastage on site when components do not fit. Designing to match the standard sizes of sheets and panels can help to avoid off-cuts, and the use of smaller board sizes can assist in dealing with complex geometries.

The carpet tile manufacturer Interface® has used product design to reduce site waste. It has created a 'i2' carpet tile that allows the tiles to be laid in any direction. This reduces cutting waste as well as making installation quicker and easier.

It is estimated that more than one million tonnes of plasterboard waste is produced in the UK each year, and wastage of 10–35% often occurs on site through wasteful design, off-cuts, damaged boards and over-ordering.[3] Waste also arises from the removal of plasterboard during refits and soft strip prior to demolition. In a WRAP case study of the Tate Modern, the use of fair-faced or rendered finishes reduced the dry-lining costs by £43,240 and avoided four tonnes of potential waste.[4]

These techniques will certainly reduce waste arising on site, but when considering a more circular economy there are additional fundamental questions about whether the materials are needed at all. In particular, can reclaimed materials be used instead, and can the need for the materials be designed-out completely?

REUSING COMPONENTS AND MATERIALS

Moving to a more circular economy does not mean that each industry has to create closed loops for its components and materials. Materials designated as waste by one industry can become a valuable resource for another; this is the principle of industrial symbiosis (as explained in Chapter 1).

Chapter 4 highlighted that the amount of materials salvaged from demolition sites is dropping, and there has been a shift away from using reclaimed materials in the UK. However, there are architects and interior designers who are embracing the opportunities to incorporate reclaimed materials into their designs.

These opportunities have to be considered right at the start of the project. The 'ICE Demolition Protocol' proposes carrying out a pre-demolition audit of existing buildings and infrastructure on the site to identify materials that could be reused in new buildings.[5] Pre-refurbishment/pre-demolition audits are now required by environmental assessment schemes such as BREEAM and the Greater London Authority's 'Circular Economy Statement Guidance'.[6] The latter requests that an independent audit is undertaken and submitted as part of a planning application.

A pre-demolition/pre-refurbishment audit is one step in the process. It will achieve little if it is not carried out in collaboration with the demolition contractors, and there must be allowances in the programme and the contract to ensure that the recommendations can be implemented.

The UK Green Building Council has published valuable guidance on reusing products and materials in built assets.[7] It proposes creating an inventory that records all the relevant details including the time required to reclaim the components, the potential storage options, and the quality and condition of each product/material.

Many of the barriers to reclamation can be bridged by engaging with third parties who specialise in reclamation of buildings, components and materials to identify the items to be salvaged for reuse. Some examples of these brokers are covered in Chapter 11.

Reclamation in practice is fraught with complications, but it can be done. When the Demolition Protocol was implemented on the London 2012 Olympic site, a wealth of materials was identified for potential reuse, and some were reused on site. Over 220 buildings had to be demolished on the site of the park along with walls, bridges and roads. The Olympic Delivery Authority set an ambitious target to reuse or recycle 90% of the material (by weight) arising through the demolition works, prioritising reuse on site over recycling. Pre-demolition audits were carried out on all the buildings and infrastructure, and the results of these audits were compared with the demand for materials for the new development. The cost of reclaiming all the materials was deemed too high, so the Olympic Delivery Authority focused on those components that could be reused locally. This was mainly the yellow stock and Staffordshire Blue bricks, the granite kerbs and setts/cobbles, roof tiles, street furniture and concrete kerbs and paving.[8]

Finding reclaimed materials and components is only part of the story. More fundamental is whether traditional models and contractual relationships can be changed to reduce waste and create a more circular outcome.

Regent's Place, 338 Euston Road, British Land – circular office fit-out

Increasingly shorter lease periods, especially in the office and retail sectors, are driving more frequent refits. In the UK, full repairing and insuring (FRI) leases require the tenants to return the space to its original, empty state, stripping out all the partitions, finishes, kitchens, tea points and other fit-out items. Once empty, the landlord will often upgrade the space to attract new tenants. And then the incoming tenant fits out the space again using new resources and materials.

British Land has recognised how the traditional leasing structure and landlord/tenant arrangements are generating waste and increasing the demand for raw materials, so for Regent's Place it has pioneered a new, circular solution.

The landlord/tenant split causes two problems: tenants will let the space get run-down towards the end of the lease as they know that it will be stripped out, and they may not have facilities staff who can maintain the building services equipment in the fit-out.

When the tenant vacated seven floors of Euston Road after five years, British Land aimed to create a circular fit-out, retaining as much as possible and rehoming anything that could not be kept. The existing fit-out that British Land inherited from the tenants was very bespoke and, according to the letting agents, there were design flaws, such as the location and size of the meeting rooms, and a lack of daylight and views out.

Also, there were no records for the additional cooling plant and other local services that were installed by the tenants. This made it difficult to retain the kit, so British Land had to assess all of the equipment to establish its condition and ensure it had the required life expectancy for another tenancy.

Paul Jaffe, Senior Asset Manager at British Land, notes that 'more and more customers want to occupy space as sustainably as possible, but operating space isn't the main reason most organisations exist; it's not their specialist area'.

He decided that the best way to resolve the landlord/tenant split was for British Land, as landlord, to retain the responsibility for the maintenance of

Figure 7.06: Regent's Place, 338 Euston Road – circular office fit-out.

the space. It would provide all the furniture, kitchen and partitions, and it would retain responsibility and would service the plant and equipment. This would ensure that the space was maintained for the long term and provide the continuity of service records. Jaffe believes that tenants prefer landlords to provide these services to save them the time and effort of finding and employing their own facilities teams.

The basic bones of the fit-out were, in fact, very good. The only issue was the rather garish orange finish to the wood. The finish was stripped off and replaced with a white satin finish.

All of the meeting room furniture was remanufactured on site using Rype Office (see case study in Chapter 12). All of the partitions were retained and, notably, all of the fan-coil units and air-conditioning units were reused, even with 23% being moved to new locations. The kitchen and furniture that was not used in the fit-out was used in the new café in the building.

The overall cost of upcycling the space was the same as the traditional approach of restoring it to Cat A (raised floors, suspended ceilings, and basic mechanical and electrical services). Seven floors were restored (see Figures 7.06 and 7.07) and two further floors were stripped back to Cat A. To corroborate Jaffe's approach, the restored floors were let immediately at a 13% rental premium.[9]

Figure 7.07: Regent's Place, 338 Euston Road – circular office fit-out.

The reuse results are impressive: 246 items of furniture were remanufactured, 2,980 kg waste was avoided and 12,145 kg CO_{2e} greenhouse gases were saved, with a further 6,060 kg CO_{2e} saved in the café fit-out.[10]

As Jaffe says, in the long term, 'as a business, we're going to get judged by the full life cycle of our buildings'.

LEAN DESIGN

A hundred years ago, one of the founders of the engineering company Brown Boveri (now ABB) said that the art of engineering is the elimination of parts.[11]

For every component that is used in a building, there will be associated waste from mining, manufacturing, maintaining and ultimately disposing of the element. Slimming down the design to the bare essentials can create a virtuous circle where both the embodied and the operational impacts of the building can be reduced, as shown in the next two case studies. The Enterprise Centre case study in Chapter 10 also shows how using lightweight, biological materials and reclaimed components can lead to a leaner design with considerably less resource use.

The Forge, Landsec – a kit of parts

The Forge, Bankside is Landsec's first net zero carbon development, offering 12,900 m² of Grade A[12] office space across two nine-storey buildings surrounding a public courtyard. It may look like any other office building, but its facade hides a series of innovations that make it an excellent example of lean design, as well as aiming to be a net zero carbon building in operation.

Neil Pennell, Head of Design Innovation and Property Solutions at Landsec, used The Forge to pioneer the first application of platform approach to design for manufacture and assembly (P-DfMA) on a live construction project. This new approach applies manufacturing techniques to buildings based on a pioneering 'kit of parts' using a standard set of components that can be combined to produce highly customised structures. The idea responds to the UK Government's 2018 challenge to the industry to transform the construction sector by applying P-DfMA.[13]

Pennell wanted to demonstrate the technique on a live project, but first he had to prove it would work, so he set up a parallel design team, who took the early designs for The Forge and applied the platform design approach. He commissioned Bryden Wood Technology (multi-disciplinary engineers) and Easi-Space (a prototyping organisation) to design and build a field trial of the concept.

The prototype platform is a hybrid steel and concrete structure (see Figure 7.08). It uses a primary steel frame based on a 9.0 m grid constructed from standardised fabricated beams and columns. The secondary structure is formed by a series of concrete beams at 3.0 m spacings running perpendicular to the steel frame. The concrete beams are formed using a Tata Steel composite floor deck profile known as ComFlor®. The profiles are inverted and fitted with reinforcement mesh to create a ComFlor beam, which is lifted into position using a specially adapted reach stacker. Reusable shutter tables are raised into position between the ComFlor beams, complete with pre-assembled reinforcement cages. Each of the 9.0 x 3.0 m bays are filled with self-compacting concrete, which can be pumped from above or below. When the concrete has reached the required strength, the temporary works elements are struck (removed) and reused on the next floor.

The composite action between the steel and concrete optimises the structural performance, reducing the amount of material needed. The concrete is placed in situ to minimise cost and transportation requirements. Although the concrete elements are not able to be fully disassembled, the temporary works used to create the concrete structure are reusable, both within the project itself and in the construction of future buildings. Most of the steelwork components have reversible, bolted connections, enabling them to be recovered and reused when the building reaches the end of its life. Similarly, the facades and MEP services modules all have bolted connections, making them easy to disassemble.

The final structure is lightweight and highly efficient, using less steel when compared with a traditionally designed steel frame building, and optimising the amount of concrete used. In the case of The Forge, 46% less steel has been used, significantly contributing to the 22% reduction in embodied

Figure 7.08: The Forge, Landsec, external render.

carbon achieved, when compared with the traditional baseline scheme. Of this reduction, 19.4% was achieved through implementing the P-DfMA solution, and the remainder through careful design and detailing.

The MEP services modules are designed to integrate with the structure, using precast fixings in the concrete slabs, reducing the need to drill the slab. The integrated services design reduces the combined structure and services zone at each floor, allowing increased-height perimeter glazing and improved daylight penetration.

The design team examined the existing approach to designing and constructing a steel and concrete building, tore up the rulebook and started again. Pennell's experience is that the typical design and construction process is ripe for improvement: the structural engineers' conceptual design goes out to tender, the contractor appoints the steel contractor who develops the detailed design,

Figure 7.09: The prototype platform design for manufacture.

and this then goes back to the structural engineers for approval, usually involving a few iterations before fabrication can proceed.

This whole process can take several months and happens later in the project, after the steelwork contractor has been appointed. The final design needs to be completed to allow other key packages to be fully coordinated with the structure. Responsibility for the design passes between the parties, often resulting in late changes with knock-on impacts on other trades, creating time and cost overruns.

Instead, for the prototype Bryden Wood Technology produced a Stage 5 BIM model where the steelwork components were designed to a level from which the steel could be fabricated. The BIM design was then translated into a fabrication software model by Easi-Space and sent directly to the steel stockist and specialist fabricator. The fabrication model data was converted into machine code for the automated cutting and fabrication processes, and accurate jigs were used for final assembly and welding. This allowed the steel to be manufactured to within a few millimetres' tolerance. The whole process took a couple of weeks, rather than months, providing further evidence of the improvements that could be achieved from adopting a P-DfMA approach:[14]

- Construction productivity improves by 55% and uses 35% fewer operatives on site.
- Cost savings are expected to reach 33% when compared to traditional construction techniques.

- Reduction in embodied carbon from material efficiencies.
- Shorter design periods through reuse of pre-designed components held in a catalogue database.
- Construction accuracy levels are improved dramatically by using multi-skilled labour teams and automated assembly processes.
- Minimises over-ordering of materials and reduces waste, further reducing carbon emissions.

Pennell notes that 'depending on the building context we can build 70–80% of the building from stuff we've already designed while still allowing design freedom for the architects'.

The Forge project is now applying at scale the principles developed in the prototype to show how a P-DfMA approach can deliver 'faster, better, cheaper, safer and greener' than traditional construction.

Building offices using kit-of-parts platform design, automated construction processes and digital technology has the potential to revolutionise construction – making it a safer and more productive process, reducing time and cost, and decreasing carbon emissions. Ultimately it will provide more and better jobs in construction, and deliver efficient sustainable spaces for office workers, helping the UK to meet its target to be net zero by 2050.

The savings achieved on this project are remarkable, and show how lean design has knock-on impacts on productivity, safety and cost savings.

CONCLUSION

The design decisions made at the start of a project will have a profound influence on the amount of waste generated. Deciding to refit or refurbish an existing building instead of demolishing and building new is often a difficult path to follow, but there is a demonstrable market for buildings with history. Similarly, using reclaimed components in design requires a more flexible approach where the design is perhaps driven by what is available instead of the original vision. Finding ways to slim down the design to reduce the demands for components has to be done by considering the whole life of the building.

Applying a manufacturing approach to construction has to be the future for new buildings, and the kit-of-parts approach demonstrably cuts carbon emissions as well as turning construction into assembly, with all the corresponding benefits. There are opportunities to reduce the demands for complex, resource-intensive building services components and internal fittings by applying excellent design principles, but the structure of the building needs to be robust enough to allow it to be adapted to new uses, as explained in the next chapter.

DESIGN FOR ADAPTABILITY

8

Almost no buildings adapt well. They're designed not to adapt; also budgeted and financed not to, constructed not to, administrated not to, maintained not to, regulated and taxed not to, even remodeled not to. But all buildings (except monuments) adapt anyway, however poorly, because the usages in and around them are changing constantly.

— Stewart Brand

It is rare for a building to remain occupied by its original owner for long, so even bespoke building designs will have been adapted to meet the needs of new users, or will have been pressed into accommodating a new way of working or a major change in technology that brings new demands to the space.

The recent changes in working driven by the pandemic mean that many workplaces and homes are being used very differently from before. Developers are recognising that the digital revolution and the changes in working practices mean workplaces that are conceived now could be obsolete by the time they are completed. And many buildings that are proving hard to adapt are being demolished when they are less than 30 years old.

There is a difference between designing a building that is simply flexible, and one that can be adapted to new uses. Addis and Schouten differentiate between flexible and adaptable buildings as follows:

- Flexible building – a building that has been designed to allow easy rearrangement of its internal fit-out and arrangement to suit the changing needs of occupants.
- Adaptable building – a building that has been designed with thought of how it might be easily altered to prolong its life, for instance by addition or contraction, to suit new uses or patterns of use.[1]

A proportion of the building stock has proved to be adaptable more by accident than by design. Georgian townhouses and Victorian warehouses are good examples of buildings that have been adapted for completely different uses in their time. Wren House in Hatton Garden, London, was built as a church in c.1670, and then became a charity school c.1696. The interior was damaged in the Second World War by incendiary bombs and had to be restored. It is now used as an office (see Figure 8.01).

However, many buildings are not adaptable, suffering from any number of limitations that mean they depreciate in value and are eventually pulled down.

OPEN BUILDING

There are movements and initiatives that have proposed ways to deliberately design buildings to be more flexible and adaptable.

'We should not try to forecast what will happen, but try to make provisions for the unforeseen,'[2] explains John Habraken, who has proposed an approach to design that he calls 'Open Building', a concept that has been widely adopted around the world. One of the aims of Open Building is to provide built environments that last because they can adjust and adapt to change.

Habraken recognised that there are several different, but related, ideas that could be considered when designing the built environment, including some thoughts that relate directly to circular economy thinking:

- The idea of distinct levels of intervention in the built environment, such as those represented by 'support' (the structure) and 'infill' (the fit-out).
- The idea that the interface between technical systems should allow the replacement of one system with another performing the same function, so replacing the building services should not mean removing all the finishes.
- The idea that the built environment is in constant transformation and change must be recognised and understood.
- The idea that the built environment is the product of an ongoing, never-ending design process, in which environment transforms part by part.[3]

As Habraken proposes: '... a strict separation of a long term "primary structure" from a short term "secondary structure" would assure better adaptation to new equipment and changing demands ... over the life time of the building'.[4] Open Building designs aim to separate the 'support' and the 'infill' in the way that the base building is a separate entity to the fit-out. The interfaces between different elements of the building have to be carefully designed to allow each part to be replaced without adversely affecting the other systems. This aligns with the concept of 'building in layers' proposed in Chapter 6. This separation of structure and the infill means that there has to be some level of redundancy

Figure 8.01: Wren House, Hatton Garden.

in the structure, and service cores have to be more generous to allow for future adaptation.

INDUSTRIALISED, FLEXIBLE AND DEMOUNTABLE

The Dutch Government applied the Open Building philosophy and developed a programme that combines standardisation, customisation and adaptability called industrial, flexible and demountable (IFD) construction. The design criteria for an IFD building include:[5]

- integration and independence of disciplines: installation, bearing structure, outer shell and interior finishing
- a completely dry building method: no pouring of concrete, mortar joints, screeds, stuccowork, sealant or polyurethane spray
- perfect modular dimensioning: a great deal of attention to drawings, prototype testing, quality system for drawings, and assembly instructions
- adjustability of all parts: bearing structure (limited), installation (practically unlimited), outer shell (limited and modular), interior finishing (practically unlimited and modular).

Many of these ideas go against the grain of conventional construction techniques, which include a wide range of composite components that are bonded together. For example, in the UK the most popular floor design for multi-storey, non-domestic buildings is the composite steel and concrete deck construction.[6]

HOW BUILDINGS LEARN

Stewart Brand's research into the way that buildings change and adapt with time is set out in his book *How Buildings Learn: What Happens to Buildings After They're Built*.[7] The book proposes some strategies that may allow buildings to be more adaptable to change:

- Loose-fit structures: spend more money and apply more effort to the structure of the building, less on the finishes and more on adjustment and maintenance.
- Scenario planning: as Brand bluntly puts it: 'All buildings are predictions. All predictions are wrong.' Using scenario planning to determine the alternative potential futures for the building will help to make the building designers think more about how the building could be used. This should help to lessen the chances that the building is so tailored to one use that it is quickly made obsolete.
- Simple plan form: Brand's case study examples show that 'the only configuration of space that grows well and subdivides well and is really

8 Design for adaptability

efficient to use is the rectangle'. Complex building forms often result in buildings that are harder to change, extend and adapt.
- Shearing layers: as discussed in Chapter 6, applying the idea that buildings have different, independent layers results in a design imperative that 'an adaptive building has to allow slippage between the differently paced systems of site, structure, skin, services, space plan and stuff.

Finally, Brand proposes that architects can mature from being artists of space to become artists of time, considering how buildings might change, and enabling that change.

The Multispace concept

3DReid created an adaptable building concept based on research into the design parameters for different types of building. The result was a set of design parameters called 'Multispace'.

The Multispace research compared different parameters for different types of buildings to see where the overlaps occurred and therefore the opportunities to design for adaptation. The key parameters examined included:

- storey height
- building proximity
- plan depth
- structural design
- cladding design
- vertical circulation, servicing and core design.

For example, a comparison of floor-to-ceiling heights showed that there was some overlap between typical standards for office, residential and hotel (bedrooms), whereas the retail buildings were outside the range (see Figure 8.02).

The research noted that the more generous storey heights allowed a wider range of servicing solutions. The proposed solution was a slim floor construction that allowed the most generous floor-to-ceiling heights without adding too much to the overall height of the building, along with a double-height ground floor that allowed for retail or for reception space.

Figure 8.03 provides a summary of the results of the research, showing a range of features that are intended to create a more adaptable building.

Chris Gregory, now of TEC Architecture, conceived Multispace while at 3DReid. The two practices prepared a research proposal that examined buildings that had successfully adapted to changes and survived while their neighbours had succumbed to the wrecking ball. One of the buildings examined was Abbey Mill in Bradford on Avon. It was built as a cloth mill in 1875, and is an impressive building located on the River Avon (see Figure 8.04). With the decline of the wool

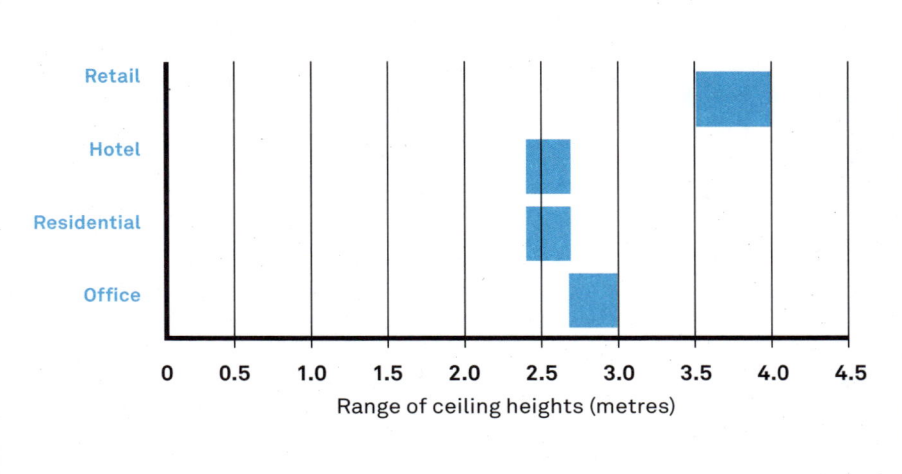

Figure 8.02: Comparison of floor-to-ceiling heights (based on the Multispace research by 3DReid).

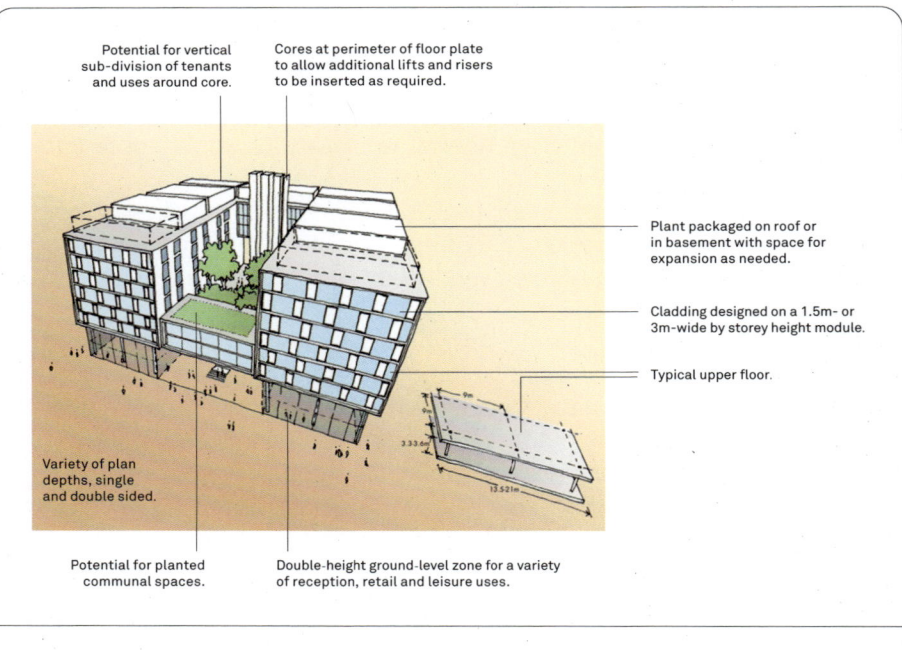

Potential for vertical sub-division of tenants and uses around core.

Cores at perimeter of floor plate to allow additional lifts and risers to be inserted as required.

Plant packaged on roof or in basement with space for expansion as needed.

Cladding designed on a 1.5m- or 3m-wide by storey height module.

Typical upper floor.

Variety of plan depths, single and double sided.

Potential for planted communal spaces.

Double-height ground-level zone for a variety of reception, retail and leisure uses.

Figure 8.03: Summary of the Multispace concept showing adaptable design features.

industry, it was eventually sold and was used for rug-making for a few years. During the First World War it was used by the army to billet soldiers. In 1915, the mill was bought by a rubber company, which used the building for storing rubber components and for rubber production. In 1967, the building was converted into offices and a restaurant, with the exception of the ground floor. In 1996 it was converted yet again into retirement apartments.[8]

These different uses would normally be considered incompatible with each other, so how has Abbey Mill managed to survive and adapt in the way that it has? TEC Architecture and 3DReid proposed seven insights into why the building had endured:

1. People like it. The building is a simple, elegant structure which dominates the riverside in the centre of town. It is taller than many surrounding buildings but people feel that it forms an important part of their townscape. They want to keep it and therefore they will continue to repair it and find use for it. The power of good architecture should not be underestimated in adaptable design.
2. There is a generous courtyard space associated with the building which acts as a social focus, a service access and a space for expansion. The space outside a building is arguably more important that the space within.
3. The upper floors of the main building are flexible rectangular spaces punctuated only by two rows of columns. They can accommodate almost anything from a production process to a highly partitioned residential care home.
4. The ground floor is deep, closely following the edge of the street and has the potential to take any number of special facilities that would not fit into the standard upper floorplates.
5. The storey height is generous: about 3.2–3.3 m. This gives plenty of room for additional services to be added if required. But, more importantly, most of the space can be naturally daylit and ventilated, which minimises the reliance on services in the first place.
6. The windows are tall, admitting daylight deep into the floor, and regular, at about 3 m centres, so there are no 'black spots' which would limit use.
7. The main original staircase is treated as a separate element that serves the main floor spaces without interrupting them. This has allowed a lift shaft to be added and a further fire escape on the far side.

The message is that a simple, robust and elegant structure will endure.[9]

The research came in useful when a volatile London property market drove the developers of Grosvenor Dock to hedge their bets and obtain planning consent for three different uses in one of the buildings. 3DReid developed the planning application for a five-storey building on the waterfront that could

Figure 8.04: Abbey Mill, Bradford on Avon, a building that has proven to be very adaptable.

be residential, office or restaurant/retail use. It was completed in 2007 as residential accommodation (see Figure 8.05).

The building includes a generous floor-to-ceiling height. As Chris Gregory notes: 'A loose fit for one building may be just right for another one; the higher ceiling heights required for offices would drive higher ceilings in residential buildings. This would increase costs, but may also add value to the residential buildings, as well as providing the building owners with the ability to adapt the buildings to new uses in the future.'[10]

The core is carefully positioned to allow for an open-plan office floorplate or flats, while avoiding creating dead-end corridors. The design includes 'soft spots' – areas within the post-tensioned slab where new service risers can be added. The soft spots are based on the principle that there should be no more than 30 m between the cores to limit the length of the horizontal service runs. Post-tensioned concrete slabs are used, as they provide a slim slab depth with no projecting beams and generous spans. This retains the much-prized floor-to-ceiling heights, while keeping the overall height of the building within the parameters of the masterplan. This construction also allows space for soft spots, as the post-tensioning tendons are typically between 900 and 1500 mm apart.

Figure 8.05: Grosvenor Dock, London, designed for adaptation.

Scenario modelling

Using scenario planning to determine the alternative potential futures for the building will make the building designers think more about how the building could be used. This should lessen the chances of the building being so tailored to one use that it quickly becomes obsolete.

AECOM uses scenario modelling to test alternative futures for buildings it is designing to explore how they can be adapted to new uses. The modelling includes exploring different grid sizes, benchmarking floor-to-floor heights, and developing building services strategies that ensure short-term flexibility and long-term adaptability. Its multi-disciplinary design teams test floorplates to see how well they accommodate different uses such as healthcare, residential, hotel and office uses, and develop facades that have interchangeable solid, transparent and louvred elements, so the facade can flex depending on the use in the space. For example, louvres for on-floor plant rooms can be exchanged for glazed elements if the space becomes an occupied area.

Of course, only time will tell whether a design is truly adaptable, as the next case study demonstrates.

The Herman Miller Building becomes the Bath School of Art

'Architects in the future must design their buildings so that they can easily be changed, either by themselves or by others. Buildings in the future should be valued as a resource.'[11] Time has shown this to be a prophetic comment from Sir Nicholas Grimshaw in 1978. He predicted the circular economy principles of adaptability and valuing buildings as a resource, and the adaptation of the Herman Miller building into the Bath School of Art demonstrates these ideas in practice.

The Herman Miller building, completed in 1977, was based on a shared vision between Sir Nicholas Grimshaw and Max De Pree, Herman Miller's managing director. The client brief – a Statement of Expectations – was more of a poem than prose, and stated that the goal was to create an environment that 'changes with grace' and 'has flexibility, is non-precious and non-monumental'. It also stated that:

Our needs will change.
The scale of the operation will change.
Things about us will change.
We will change.[12]

To respond to the brief, the original design of the factory was a double-height space formed from a 10 x 20 m steel grid, with only lightweight mezzanines and partitions, that could be reconfigured with minimal structural works or design input.

The design included a modular facade system made up of interchangeable solid, glass, louvred and door panels. Herman Miller used this feature several times during the building life, allowing the building to be adapted to house office and R&D spaces as well as the fabrication areas.

The entrances and outdoor seating areas were created with indent bays in the frame that could be moved or removed as required. The factory was serviced by two overhead gantries that ran the full length of the building, allowing new connections to be added and dropped down to the required location.

These features made the building incredibly flexible, but the real test of the original vision came when Grimshaw was appointed by Bath Spa University in 2016 to adapt it into the School of Art.

The new brief required two-storey accommodation, so the existing roof line was raised using new roof beams that span over the existing structure. The new roof incorporated more insulation and extensive rooflights to provide daylight into the deep floorplate, as illustrated in Figure 8.06. The raised roof allows space for the high-level service gantries, which are now packed with the distribution systems required to service the new building (see Figure 8.07).

A new rooftop extension was built, using pre-existing pad foundations that were delivered in the original design, to allow for future expansion. The external facade panels were refurbished and the existing glazing was replaced with double-glazed units throughout. Figure 8.08 shows the new cladding, one of the indent bays and the roof extension.

Figure 8.06: The conversion of the Herman Miller factory into the Bath School of Art.

Figure 8.07: The Bath School of Art interior showing service gantry.

Figure 8.08: The Bath School of Art, showing the modular cladding system and the extension on the roof.

By retaining many of the original features and ensuring that new elements follow the same ethos of adaptability, the new incarnation aims to retain its ability to adapt in future.

This building has surpassed its original brief by not only being flexible enough to accommodate the change demanded by the manufacturing industry for the last 40 years, but by successfully morphing into a higher education building, while retaining its essential character. In echoes of Winston Churchill's comment that 'we shape our buildings; thereafter they shape us', the Head of the School of Art at Bath Spa University observes that 'In the past we had to adapt our practices to suit the building, but now we have a building that doesn't dictate, but informs how we can teach differently'.[13]

CONCLUSION

The research and case studies show that it is possible to design buildings that are more adaptable. Whether they will be successfully adapted depends on economic and social factors as much as technical design-related parameters.

The use of a layered approach allows buildings to be flexed and adapted more readily. In particular, a separation between the primary structure, the facades, the services and the interiors of the building allows the structure to be retained while the facade is replaced, or the interiors to be changed into new layouts while not being dictated by structural walls in awkward locations. Equally, the building services need to be as accessible as possible and not entangled with the structure or encased within walls, floors or behind finishes.

Other factors that would help to make the building more open to adaptation are:

- over-engineering the structure and foundations to accommodate changes in use and the potential expansion of the buildings
- exploring different structural solutions that provide clear floorplates and options to add or remove parts of the floorplate or to provide penetrations through the floors for different servicing strategies
- using a simple plan form to allow reconfiguration of internal spaces and expansion
- positioning of cores and generous sizing of risers to allow for future changes in use
- scenario modelling to show how the proposed buildings can be adapted to different uses – this will help to inform the optimum floor-to-ceiling heights, service zones, floorplates and structural grids that can accommodate different uses
- generous floor-to-ceiling heights with abundant levels of daylight are a feature of many buildings that have been through several adaptations, as is the provision of external space
- applying the principles of design for disassembly (see Chapter 9) should make the building more adaptable, as the building components can be separated more easily and different elements will be more accessible for repair or replacement.

Building-in the ability for interiors and local services to adapt to new uses is potentially wasteful, as it is likely that these elements will be stripped out and replaced if the building is adapted to a new use. Designing-in some degree of flexibility to allow internal partitions to be relocated and local services to be extended or augmented may be appropriate, depending on the building use and the lifespan of the components.

Designing simple, robust and elegant buildings that are valued by people helps ensure that buildings endure. Alternatively, buildings can be designed as modular structures that can be demounted and reconfigured, as explored in the next chapter.

DESIGN FOR DISASSEMBLY AND REUSE

At its core, a circular economy aims to design out waste. Waste does not exist: products are designed and optimized for a cycle of disassembly and reuse.

— World Economic Forum

Design for disassembly

Being able to extract components from buildings intact is an important part of applying the circular economy. Components that have reached their end of life or need repair have to be easily accessible, rather than buried in other materials. Elements of a building that are to be removed as part of a refurbishment, refit or demolition have to be recovered without damage so that they can be reused, remanufactured or recycled.

Materials that are toxic have to be isolated and returned to industrial processes. Biological materials have to be separated and uncontaminated so that they can be composted or used to generate biogas.

To achieve these aims, buildings have to be designed with a thought for what happens to their constituent parts at the end of life.

Designers will often consider plant replacement strategies as part of the design, and will work up drawings showing how large components such as chillers can be removed and replaced. If the brief demands, designers may also prepare strategies showing how cladding panels can be replaced or internal partitions can be relocated. It is conceivable, though rarely done, to have a strategy for reclaiming components and materials at end of life, and to design to enable disassembly of the building.

This chapter summarises the principles for deconstruction and includes some case studies showing how a few designers are thinking about the end of life of buildings and how they can be deconstructed and used again. Chapter 12 sets out some different business models that turn buildings into materials banks for future generations.

PRINCIPLES OF DESIGNING FOR DECONSTRUCTION

There is a wealth of research about designing buildings for deconstruction (or disassembly), and various guides that set out principles to be followed. Notably, a guide by the CIRIA[1] sets out principles (Table 9.01) and design guidance, by building element, on designing for deconstruction.

A survey of demolition contractors[2] showed that some of these principles are considered more beneficial to deconstruction than others, namely:

- having mechanical and reversible (e.g. not chemical) connections
- ease of access to connections
- independent and easily separable elements of the building, e.g. structure, envelope, services and internal finishes
- no resins, adhesives or coatings on the elements.

The survey also identified the need for information about the building, including full as-built drawings and a deconstruction plan to help the demolition contractor understand how to disassemble the building.

Design principle	Component reuse	Component manufacture	Material recycling
Provide identification of materials and components	✓	✓	✓
Provide guidance for deconstruction	✓	✓	✓
Design for simultaneous, parallel disassembly and deconstruction	✓	✓	✗
Design for deconstruction using common tools and equipment rather than bespoke tools	✓	✓	✗
Minimise the number of different types of components	✓	✓	✗
Mechanical in preference to chemical connections	✓	✓	✗
Consider using modular construction	✓	✓	✗
Provide good access for deconstruction, especially connections	✓	✓	✗
Design components sized to suit appropriate means of handling	✓	✓	✗
Provide adequate tolerances for assembly and deconstruction	✗	✓	✗
Design connectors, fixings and components for repeated use	✗	✗	✗
Consider using standard grids	✗	✗	✗
Use the minimum number of different types of connectors	✗	✓	✗
Use the minimum number of interfaces and connectors	✓	✓	✗
Consider the use of prefabrication	✓	✓	✗
Minimise the number of different types of materials	✓	✓	✓
Use alternatives to toxic and hazardous materials	✓	✓	✓
Make inseparable sub-assemblies from the same material	✓	✓	✓
Eliminate the use of secondary finishes to materials	✓	✓	✓

✓	Very important
✗	Less important

Table 9.01: Design principles for deconstruction for different end of life outcomes. Courtesy CIRIA.[3]

Doctoral research by Paola Sassi into 'closed loop material cycle construction' proposes a set of technical criteria that can be used to determine whether a material or component can be disassembled.[4] This research is based on an extensive review of literature and analysis of the essential criteria. Table 9.02 shows a summary of the results and examples of compliant and non-compliant components.

Table 9.02 can be used by designers to test out whether their proposals can be deconstructed at end of life.

The research by Sassi proposes the following:

- Access to building components and to fixings is essential. Embedding elements within other elements makes it impossible to separate them without damaging one or both of the components. Similarly, fixings need to be easily accessible and not require excessive force to undo them. For example, a fixing cast in concrete would be inaccessible.
- Loose connections and friction fittings are inherently demountable, with mechanical fixings having a relatively high demountability and chemical fixings being the least demountable.
- A fixing should not contaminate the elements. For example, a rusted metal fixing in timber may break off and remain embedded in the wood. On the other hand, organic polymer adhesives can be burned off metal sections when being melted down for recycling without affecting the quality of the new steel.
- Specifying mechanical fixings is not sufficient to enable disassembly; the fixings also have to be durable. For example, using screws that corrode quickly will not facilitate disassembly.

The CIRIA report notes that: 'The value of a building element at deconstruction will reflect its condition at that time, the work needed to refurbish it for later reuse and the level of performance that it might be able to achieve.' Therefore, components or elements need to be easily cleaned, maintained and serviced. For reuse after deconstruction, the elements have to be removed from the building with as little damage as possible. To enable reuse, the components have to be: not obsolete, not corroded or worn and compatible with other elements.

DIFFICULT DEMOLITION WASTES

Research by BRE in collaboration with the National Federation of Demolition Contractors identifies materials that are currently being used that are difficult to reclaim or recycle at the demolition stage.[5] Table 9.03 shows some of the products identified in the report that are currently entering the waste stream, along with the recovery issues and potential opportunities for recovery.

Table 9.04 shows some of the products that the BRE report suggests will be entering the waste stream in the future.

Process	Requirement	Compliant example	Non-compliant example
Ability to access	All components are readily accessible and removable	Door	Service duct embedded in concrete
Accessibility of fixings	All interfaces/connection points, fixings are identifiable and accessible	Screw-fixed timber cladding	Plasterboard fixings under full coat of plaster
Types of fixings	Fixings have: • the ability to be removed (mechanically, or with solvents) from the element or alternatively integrated within the recycling process • the ability to ensure a high percentage of recovery of the material • the ability to ensure a high quality of the material recycled without contaminants • the ability to remain operational long term.	• Water-based/ soft adhesive • Screw/bolt	• Insoluble adhesive • Rivet
Durability of fixings	Design joints will remain operational and do not compromise removal of element over time	Stainless steel fixing	Steel fixing that may rust
Information	Sufficient information is provided OR no information is required to enable dismantling	Brick wall	Proprietary temporary building

Table 9.02: Summary of technical criteria for design for deconstruction.

The BRE report shows that elements that are bonded together or contain hazardous materials create problems at end of life, based on current demolition practices. Table 9.03 illustrates the legacy left by the previous generation of buildings and the issues that were not considered when the buildings were designed and constructed. Table 9.04 provides some examples of components that could create new problems when the next generation of buildings is demolished. The list includes components that are designed to improve the efficiency of the building in operation. Even elements that are jointed with mechanical fixings, such as the composite floor cassettes and structural steel sections, are noted in the report as being difficult to handle using mechanical demolition.

Product	Recovery issues	Recovery opportunities
Aerated concrete blocks	Bonded with mortar Not suitable for recycling into concrete aggregate	Can be used as a substrate for green roofs
Asphalt roofing	Classed as hazardous waste Bonded to roof structure	
Extruded polystyrene foam (XPS)*	Older board may contain ozone-depleting substances (hazardous waste) Floor insulation is encased in concrete	
Laminated wood	Not readily recyclable	Potential to be composted Energy recovery
Metal insulated panels	Metal-insulated panels are typically recycled for the metal Insulation is stripped out, and typically is sent to landfill	Post-2004 panels taken out of buildings can be reused (as these will not contain ozone-depleting substances)
Phenolic foam boards	Mostly landfilled	Opportunity to incinerate
Phenolic pipe sections	Usually sent for disposal as hazardous waste at a landfill	Opportunity exists to incinerate Potential for mechanical recycling
Polymer composites	Typically landfilled in the UK	Several recycling options have been developed, including the reintroduction of ground fibre-reinforced polymer waste into the production process
Polyurethane rigid foam (PUR)	Boards containing ozone-depleting substances (e.g. pre-2004) need to be disposed of as hazardous waste	New approaches to reuse and recycling
Wood-based board	Not readily recyclable due to additives and adhesives	New technology is being developed to recycle these products, using microwaves and moisture to burst the bonds in the board without damaging the fibres

Table 9.03: Examples of difficult demolition wastes currently entering the waste stream. Summarised from BRE, *Dealing with Difficult Demolition Wastes*.

This is corroborated by research by the University of Cambridge, which shows that even structural steel beams with bolted connections that could be reclaimed are typically cut into sections using hydraulic shears.[6] The additional time taken to unbolt the complex joints will not provide sufficient return to justify the extra labour required. The research goes on to propose that the use of novel

Product	Recovery issues
Brick slips/brick tile system	Bonded to composite insulation panels, so difficult to separate
Closed-panel timber frame: phenolic foam applied in factory	Time-consuming to separate, mineral wool and phenolic foams generally end up in landfill
Concrete with macro fibres (steel fibres, plastic fibres)	Not known whether this material can be recycled back into aggregates
Floor cassettes (composite)	Difficult to segregate materials on site
Insulated plasterboard	Difficult to segregate on site
Insulating concrete formwork	Difficult to segregate on site, risking contamination of recyclable material
Phase-change materials (PCMs)	It is not currently known how PCMs would affect recycling
Roof cassettes	Slow to dismantle and segregate materials

Table 9.04: Examples of difficult demolition wastes in the future. Summarised from BRE, *Dealing with Difficult Demolition Wastes*.

jointing techniques such as Quicon, ATLSS and ConXtech could reduce the time taken to facilitate greater reuse in the future.

Flooring systems are one of the most intractable problems when designing for deconstruction, as they are often composite systems using steel and concrete to span large distances.

Demountable composite flooring
Composite metal deck flooring is one of the most widely used and efficient floor systems in multi-storey steel-frame buildings. These composite flooring systems consist of a profiled metal deck with a poured concrete topping, supported by steel beams. Composite action between the steel beams and the concrete floor is created by welded studs. These connectors are welded through the steel decking to the steel beams and cast into the concrete, making deconstruction of the composite floor system impossible.

Previous researchers have proposed replacing the welded shear stud with a bolted shear connection to allow the slab to be separated from the beam, but because of a lack of knowledge of the structural behaviour of bolted shear connections and absence of design guides, it has rarely been used.

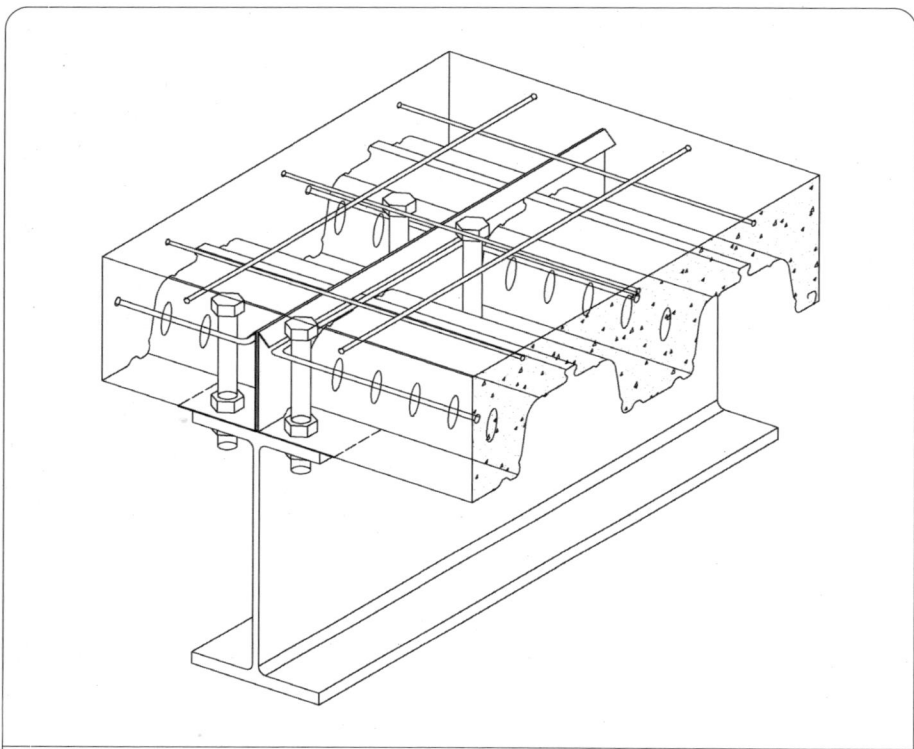

Figure 9.01: Composite metal deck flooring with demountable, bolted connections instead of welded shear studs.

Dennis Lam and his team at the University of Bradford have proposed two enhancements to this concept that allow both the steel beams and the concrete slabs to be reused. And like many of the best ideas, they are elegantly simple.

First, instead of completely replacing the studs, they have proposed using the traditional shear studs, but introducing some threads near the end of the studs with bolted connections to the beams (see Figure 9.01). This retains the same material properties and the composite action provided by the shear studs but allows the composite floor to be unbolted and disconnected from the supporting beams.

Second, they have proposed the idea that the concrete slabs can be salvaged by cutting them into segments for reuse as precast concrete units (as shown in Figure 9.02).

This allows the potential reuse of both the steel beams and the concrete slabs, and even if the slab is discarded, the system allows the steel beams to be reused.

The team at the University of Bradford have tested this approach using full-scale composite beam tests, and demonstrated that the performance of the reused flooring system is comparable with the original use. They have also contributed to a Steel Construction Institute publication that provides design guidance and criteria on the use of the flooring system using bolted shear connectors.

This is a highly scalable innovation that could be implemented across the world to provide future generations with buildings that have the capacity to be demounted.

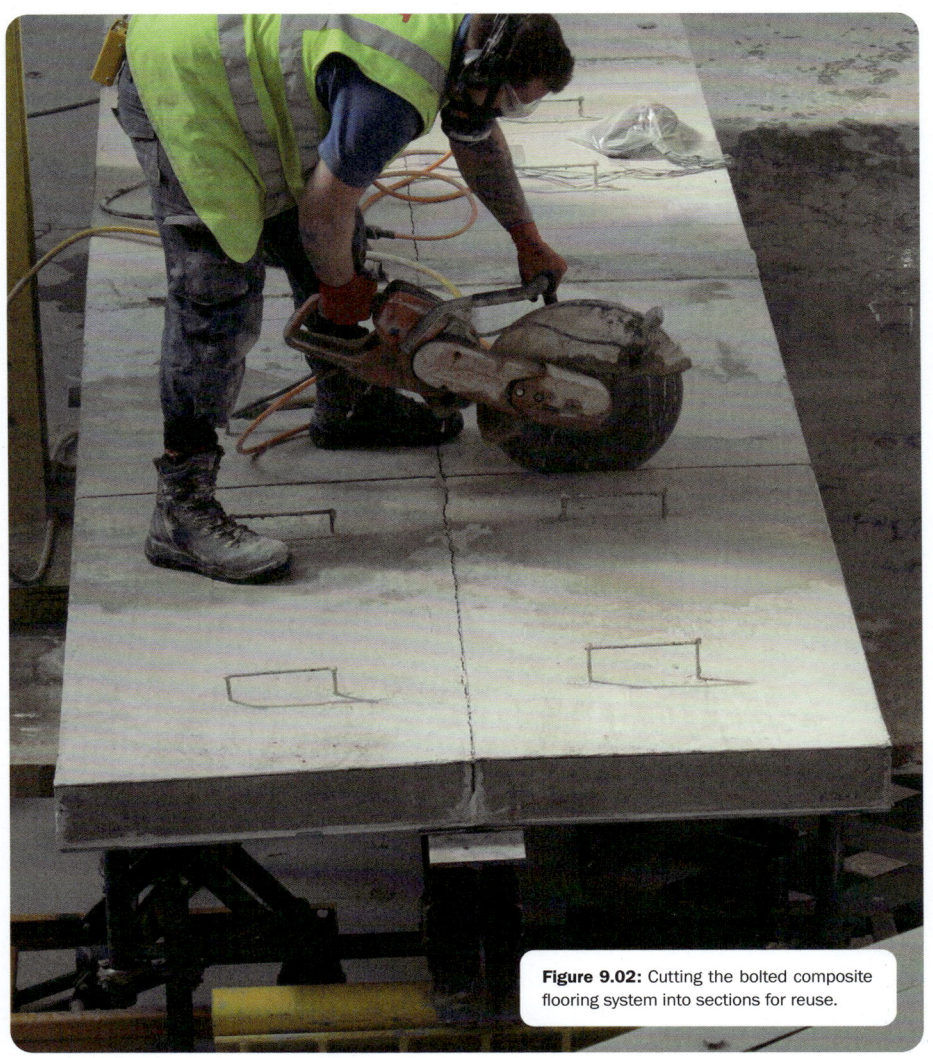

Figure 9.02: Cutting the bolted composite flooring system into sections for reuse.

Offsite construction and design for manufacture and assembly

Offsite/modular construction and design for manufacture and assembly (DfMA) is the future of the industry. Applying a manufacturing approach to buildings will be far more efficient and effective, as demonstrated by the case studies in this book (notably The Forge in Chapter 7). This approach will create buildings that use resources far more efficiently, avoid waste on site, and are easier to disassemble than previously. However, there is a real risk that some types of modular construction could work against the circular economy by using composite components and exacerbating the problems with demolition and disposal.

Robin Powell, a student in Demolition Management, was inspired to research[7] the demolition of offsite construction when he heard about the 'premature' demolition of a high-rise, modern-methods-of-construction (MMC) hybrid-designed student accommodation. The buildings were demolished after only ten years to make way for a new infrastructure project. Powell notes that 'some of the walls were constructed of decorated plasterboard, stuck to a cement based, wood chip containing fire resistant board, stuck to "Styrofoam" insulation, stuck to a vapor proof membrane, stuck to "Styrofoam", stuck to a fire resistant board, stuck to plasterboard. It was found to be "unrealistic" to separate this amalgam of materials into their constituent parts, resulting in the "mixture" being sent to landfill at great expense'.

This example shows the difficulties that arise when the construction techniques are new and unknown; the demolition contractors had not allowed for the sheer volumes of unseparable, unrecyclable waste that arose from the demolition. It also highlights a potential risk of MMC if the products that are built do not consider circular economy principles at the design stage.

Perhaps the best way to ensure design for disassembly would be to involve the demolition industry in the design process to highlight potential issues at the outset.

DESIGNED FOR DISASSEMBLY

There are some examples of buildings that have been designed for disassembly. The Prologis distribution facilities building near Heathrow was fabricated with a two-bay portal frame, and all the steel members are stamped with the section size and grade to enable them to be identified and reused. Approximately 80% of the portal frame structure is designed to be reusable, along with 95% of the floor beams and 100% of the galvanised steel components (SteelConstruction. info). Portal frame buildings are one of the few types of construction that are dismantled and reused, as there is a market for secondhand portal frames for use in the agricultural industry.

Some designers have taken the idea of designing for deconstruction to heart and have created buildings that can be disassembled and recycled with relative ease, as shown by the first case study from Werner Sobek. The second case study explains how XXarchitecten has created a 'building' from modular components that is now in its third incarnation, having had two previous lives on two different sites. The last case study shows how AHMM has created a reconfigurable partitioning system for Google, illustrating how design for disassembly and adaptability are often two sides of the same coin.

WERNER SOBEK: URBAN MINING AND RECYCLING EXPERIMENTAL UNIT

In Germany, Werner Sobek has long been an advocate of environmentally sustainable buildings, and has designed prototype houses that are highly energy efficient and that consider the end-of-life recyclability of the components and materials.

Werner Sobek is involved in the Urban Mining and Recycling Experimental Unit (UMAR), part of the four-storey research building on the campus of the Swiss Federal Laboratories for Materials Science and Technology (Empa) in the municipality of Dübendorf, Switzerland (see Figure 9.03).

The aim is to demonstrate how to create a structure that is fully reusable, recyclable or compostable. Most of the materials have been drawn from existing technical and natural cycles, and the dry connections ensure that they can be separated and returned to the cycles at a later date.

The supporting structure and large parts of the facade are made from untreated wood, and the facade also uses aluminium and copper, which can

Figure 9.03: Urban Mining and Recycling Experimental Unit (UMAR).

be cleanly separated for recycling. The copper sheets for the facade were painstakingly arranged according to their colour and were reclaimed from other buildings, including a church.

All of the building's structural bonds enable it to be fully dismantled and separated into its individual constituent parts after an expected lifespan of five years. The designers employed traditional techniques such as trims and fasteners to connect the façade, and avoided adhesive joints in such distinctly tricky areas as the vapour barriers and the waterproof seals in the bathrooms. Press fittings were used instead of silicone joints, and screws and hooks were used to hold components together.

The MycoFoam insulating panels (manufactured by Ecovative Design) are made from wood chips and agricultural waste and look like chipboard, but with the fibres bonded together by fungus-based cultures that act as a natural adhesive.

As Sobek notes,[8] 'A cycle has no beginning and no end. Accordingly, the products and materials used in the building occupy a wide range of different positions in their respective cycles. Some are new or as good as new, while others have already been reused or recycled once or several times before. The only thing that matters, however, is that they can be returned to their cycles when the unit is dismantled'.

XXARCHITECTEN, ROTTERDAM

XXarchitecten has an impressive track record in designing demountable buildings for a limited lifespan that can be redeployed for different purposes or fully disassembled, allowing the materials and components to return to the biosphere or be reused.

Villa Camera is a rare example of an industrial, flexible and demountable (IFD) building (see Chapter 8 for an explanation of an IFD building), whose components have actually been demounted and reused as three different buildings.

Villa Zebra, Rotterdam

In its first life, the building was the Children's Hall of Art in Rotterdam. Completed in 2001, the aim of Villa Zebra was to create a cultural workspace for children that included exhibitions, cooking areas, a children's café, a theatre and studio spaces. The building was designed to have a five-year life with the ability to be extended or decreased in size, depending on the success of the venture.

XXarchitecten designed a flexible, demountable building from modular steel units (6 x 3 x 3 m) with mechanical connections between the units and prefabricated cladding panels (see Figure 9.04).

After five years on the site, it was dismantled and the components were stored.

Figure 9.04: Children's Hall of Art in Rotterdam.

Villa Nutcracker, Hoogvliet

Two years later the same elements were used in a different combination to build temporary accommodation for a school in Hoogvliet. Villa Nutcracker was intended to last for three to five years, until a new school building was constructed. This time, the building was made 3 m wider by adding additional units to enable the classrooms to be interconnected. Villa Nutcracker contained nine classrooms, common areas and a library.

Villa Camera, Hilversum

Then, in a remarkable third reincarnation, Villa Nutcracker was dismantled and rebuilt into three new buildings. A section of the lower part of the building has become a small school building, while another section has been added to an existing harbour building for Codarts, a school for artists in Rotterdam. Part of the building has been reassembled as Villa Camera (see Figure 9.05). Designed for Facility House Broadcast Group, the building is located in the Media Park in Hilversum and is used to store camera equipment for hire along with associated office space. Opened in 2013, it has been nominated for an architectural award because of the combination of expressive architecture and nature.

As Jouke Post says: 'It's almost a new type of building: a nice building made of boring boxes.'[9]

It is not always conceivable to design a whole building to be demountable and relocatable, but it is certainly possible to make the shorter-life elements of the building easier to change.

Figure 9.05: Villa Camera Facility House, Media Park, Hilversum.

RELOCATABLE PARTITIONS IN BUILDING INTERIORS

With building interiors having progressively shorter lives and rapidly changing working practices requiring highly flexible spaces, the idea of relocatable partitions makes complete sense. However, such systems are often bespoke and expensive, and require highly engineered designs and expertise to reconfigure.

For 6 Pancras Square, in 2016 the Google Real Estate team commissioned AHMM architects to develop a modular system that could be reconfigured in a few hours[10] without resorting to smashing down old partitions with sledgehammers and installing new plasterboard stud walls.

The system, created by AHMM and christened 'Jack', comprises simple components that can be assembled, arranged, disassembled and rearranged in a variety of ways.

Jack consists of a series of plywood 'cassettes' that form the building blocks, from which a range of configurations can be created ranging from small meeting rooms and video-conferencing booths to mobile dividers and vertical storage panels[11] (see Figure 9.06).

Different surfaces can be added to improve acoustic performance, provide privacy or give the unit a specific colour and identity. For larger meeting rooms, a split unit condenser can be added to the roof, if required, and connected to a soffit-mounted air-conditioning system.

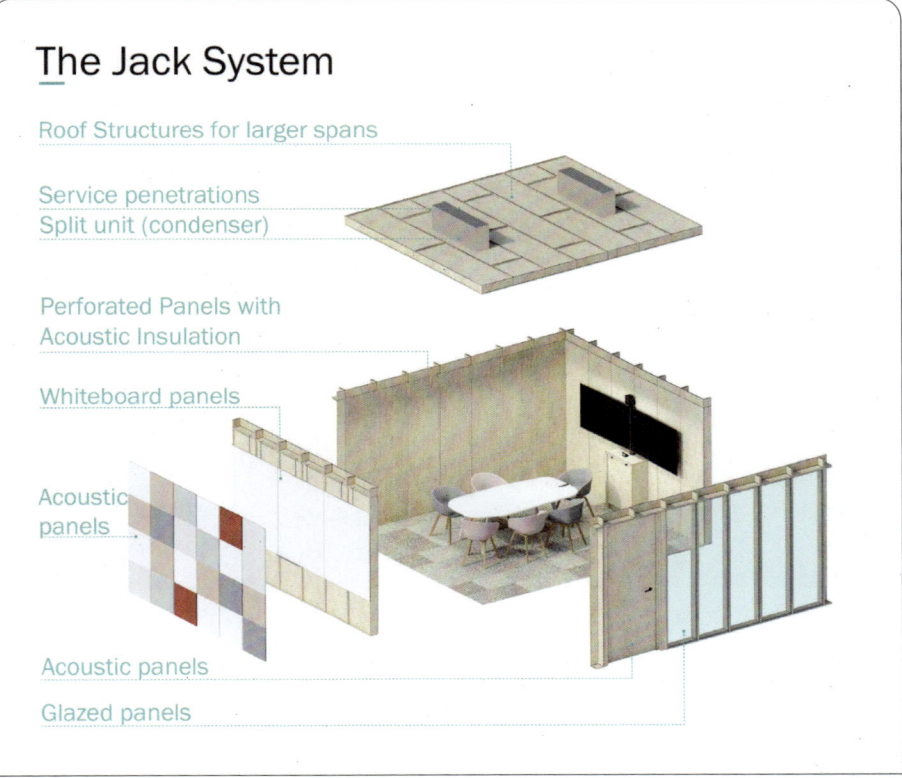

The Jack System

Roof Structures for larger spans

Service penetrations
Split unit (condenser)

Perforated Panels with
Acoustic Insulation

Whiteboard panels

Acoustic
panels

Acoustic panels

Glazed panels

Figure 9.06: Jack modular partition system design by AHMM for Google.

A total of 160 Jacks are being installed in 6 Pancras Square, and prototypes have already been installed in other international Google offices, including Dublin, Munich and Tel Aviv.

The system has also been adapted and used in the White Collar Factory (see Chapter 7).

Anecdotal evidence tells that Googlers liked using the first Jack prototypes due to its acoustic privacy, the easy integration of video conferencing and the use of natural materials.[12]

CONCLUSION

Modular, demountable buildings can be deployed temporarily on sites that are earmarked for redevelopment.

Designing for deconstruction is an essential piece of the circular economy puzzle. It allows buildings to become a kit of parts that can be reconfigured

Figure 9.07: Jack demountable partition system installed.

during operation, providing access to different layers and elements. It enables buildings to be deconstructed rather than demolished and, when done correctly, it should make the process safer and more efficient. It allows components and materials to be reclaimed intact during renovation or demolition. If this is combined with a new infrastructure, mechanisms to make buildings into materials banks and inventories that allow materials to be pre-sold before the building is even demolished, then there is the potential to close the loop. These ideas are explored further in Chapter 11.

As well as creating mechanisms to encourage salvage and reuse, the materials and components have to be selected to enable reclamation, remanufacture, recycling or composting. This is discussed in the next chapter.

10 SELECTING MATERIALS AND PRODUCTS

Buildings are gold mines of materials just waiting to be harvested.

— Ellen MacArthur Foundation

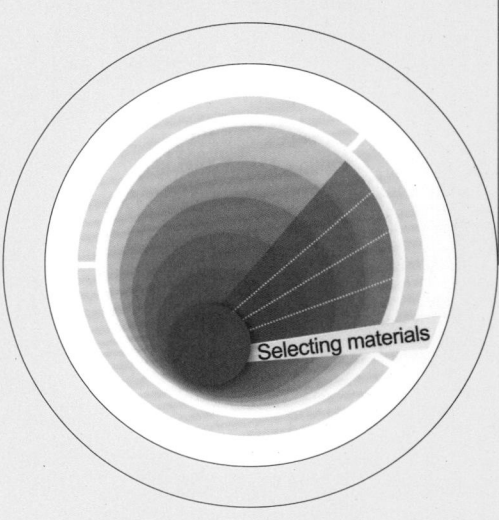

Selecting materials

In a circular economy, each material in the building and its components has to be declared and defined, the composition has to be as pure as possible, and it has to separable from other materials. This ensures that buildings can become materials banks (see Chapter 11), where materials are effectively stored for future use, rather than consumed and lost.

One of the defining principles of a circular economy is the distinct split between 'biological materials' and 'technical materials', as explained in Chapter 1. Biological materials, such as wood and sand, are used in products that are free of contaminants and toxins, so they can be returned to the biosphere at end of life. Technical materials, such as metals and plastics, are retained within industrial loops that ensure they are not lost to the economy or returned to the environment. These technical components should be reused as far as possible and the constituent materials recycled as a last resort.

MATCHING LIFETIME TO MATERIAL SELECTION

The idea of 'building in layers' (proposed in Chapter 6) aims to differentiate between components with a shorter lifespan (e.g. carpets) and those with a longer lifespan (building structure). The likely lifespan of each component should be carefully considered and the materials selected accordingly. The lifespan of internal fixtures and fittings is often over-estimated and designed for a long lifespan that is not achieved, leading to significant waste. Components that are likely to have a shorter expected lifespan can either be made from biological materials that can be returned to the biosphere or designed to be readily returned to the manufacturer for reuse, remanufacture or recycling. Components with a longer lifespan, such as the structure and fabric, should be designed to be durable and resilient, while ensuring that they can be maintained, upgraded or disassembled, as required.

BIOLOGICAL MATERIALS

Biological materials are those that can be returned to the biosphere and allowed to biodegrade. The circular economy model proposes that the use of biological materials can be 'cascaded' through various uses, rather than used just once. An example of cascading might be where solid timber is used in a building and then chipped for use in panel products.

Biodegradable materials are those that can be broken down by microorganisms in the biosphere. Sassi proposes that biodegradable materials can be grouped into four categories:[1]

1. Natural materials that can be used following minimal processing (e.g. timber, bamboo, cork, hemp).

2. Natural materials bonded with a resin or mesh (e.g. clay, hemp and straw mixtures for external walls, strawboard for internal partitions, jute carpet backing, linseed oil and natural resin to make linoleum).
3. Natural compounds used in manufacturing products, including adhesives and other polymers (e.g. natural protein to manufacture biodegradable plastics).
4. Biodegradable synthetic materials (biodegradable plastics).

Materials in the first category need only to be shredded to allow them to be composted. For materials in the second category, the aim would be to ensure that biodegradable material is not bonded with non-biodegradable materials (e.g. natural fibres used in concrete or cement products or chemical, non-biological bonding agents), as this will inhibit its ability to be returned to the biosphere. Non-toxic bonding agents can be used in minimal amounts, but this would restrict the potential uses of the resulting compost.[2]

Natural biodegradable plastics can be made from polymers such as cellulose, starch, protein and sugar molasses extracted from plants.[3] Biodegradable synthetic plastics have been developed and have some limited use for disposable packaging and plastic bags.

Substituting biodegradable materials for technical materials is one way to reduce the end-of-life impact of components, particularly those with a short lifespan or those that currently have poor reclaim or recycling rates. For example, plasterboard is currently made from a technical material that is often not returned back to an industrial process at end of life due to contamination with paint and other materials. Products such as ECOR Panel Products and Adaptavate's Breathaboard represent an opportunity to substitute a biological material for a technical one, providing multiple benefits in the process, as illustrated in the following case studies.

GatorDuct – cardboard ductwork

Building services typically have a shorter life than the buildings that house them. Systems and components that service floors and occupied areas are often stripped out as part of a refit, particularly in office and retail environments. Perhaps in these situations, the local services could be considered as consumables. So ductwork could be made of readily recyclable materials instead of using durable, engineered metals and plastics. There is ductwork on the market made from fabric and even from cardboard.

Tri-wall cardboard ductwork has been developed by GatorDuct, which has a coating made from a water-based solution with a water dispersal polymer, fire-retardant minerals and a final hydrophobic finish (see Figure 10.01). The coating can be recycled with other water- and oil-based printed paper. The cardboard is high strength and can be used by recyclers for other substantial cardboard products. Cardboard ductwork requires less insulation than steel

Figure 10.01: Cardboard ductwork.

ductwork as it has some insulating properties. It is also considerably lighter, making it easier to handle and install.[4]

Architype and the Enterprise Centre

Biological materials can be used to create lighter construction that uses fewer resources, as well as providing materials that can be returned to the biosphere.

Architype's desire for simplified designs can be traced back to its roots in self-build housing. Designing for lay builders means everything has to be as simple and pared back as possible. The practice has carried this philosophy through its projects and has designed buildings that use fewer resources through a combination of lean design and careful materials selection.

The self-build influences are evident in the lightweight timber-frame buildings that are a common theme of Architype's designs. The Coed-y-Brenin Visitor Centre in Wales exemplifies the lean, low-impact approach. Its lightweight structure uses locally-sourced sitka spruce, which is turned into structural-grade timber by using short lengths of softwood held together with hardwood dowels that swell and lock the planks together. The cladding is formed from charred local wood that means it needs no other finish. The jointing methods and lack of finishes means that there is no need for glues, adhesives or paints.

As part of the lean design philosophy, Architype champions the Passivhaus standard, an approach that aims to create comfortable buildings with dramatically lower energy demands. The standard combines high levels of insulation and airtightness with mechanical ventilation and heat recovery to reclaim any escaping heat. Cooling demand is reduced by using shading, pre-cooling the air, and using natural ventilation and night cooling strategies. Architype finds that applying these techniques on their buildings enables them to reduce heating and cooling kit to a fraction of its typical size.

All this experience came in useful when the University of East Anglia released its demanding brief for the Enterprise Centre (see Figure 10.02). The university wanted a new gateway building for its campus that could accommodate an innovation lab, a 300-seat lecture theatre, flexible workspace, teaching and learning facilities, and incubator units for new startup. And it wanted it to reduce its embodied carbon and operational carbon emissions by using innovative construction techniques and designing to the Passivhaus standard.

Focusing only on reducing embodied carbon does not necessarily fit into the circular economy ideal, as it can drive designers to substitute highly recyclable (and recycled) materials such as metals with materials with lower embodied carbon – for example thermoset plastics, which are difficult to recycle. Also, focusing on embodied carbon does not consider the other impacts associated with winning and processing the raw materials, such as scarcity or the impact on biodiversity of mining or drilling operations.

Figure 10.2: The Enterprise Centre, University of East Anglia.

For the Enterprise Centre, Architype was careful to consider all the environmental impacts over a 100-year life by consciously selecting biological materials. The designers worked closely with the contractor, Morgan Sindall, to select materials and components that were locally sourced, natural and biodegradable. This helped them to avoid the impacts of mining and transporting materials from abroad, while stimulating the local economy.

The most striking and innovative feature is the use of thatch to clad the building, which draws on the local vernacular and gives it a novel twist. The design uses prefabricated thatch panels constructed as cassettes that are slotted into place, providing additional insulation as well as acting as a rainscreen. The thatch is simple to repair or replace, and the straw can be composted at end of life. The clerestory rooflights are also thatched using traditional Norfolk reed over the pitched roof.

The timber frame building uses studwork mainly sourced from the nearby Thetford Forest, and the structural timber columns under the main entrance canopy are made from local trees turned in Suffolk. Cellulose insulation made from recycled newspaper is used as both the thermal insulation to the walls (Warmcel) and as acoustic insulation and the ceiling finish (SonaSpray).

The designers were able to reap the rewards from selecting a lightweight frame and cladding system by radically reducing the amount of foundations required. The lean design of the foundations uses a concrete floor slab constructed with aggregate from a hospital that was demolished nearby, and it uses the maximum 70% ground granulated blast slab (GGBS) cement replacement. The results are seriously impressive: the foundations and floor build-up have one-tenth of the embodied carbon of a conventional design solution, and the durable diamond ground floor finish means that replacement cycles of the floor finish are all but eradicated.

Biological materials are used for the interior finishes, including clayboard cladding with an earth finish in the atria. Additional finishes were avoided wherever possible, as tiles and carpets increase the embodied carbon, and paints, grouts and adhesives would inhibit the deconstruction of the building. The use of natural, breathable materials provides a healthier internal environment for the occupants (see Figure 10.03).

The team also salvaged components from other buildings to reduce the demands for virgin materials. The front reception desk is an unused desk from the Sainsbury Centre for the Visual Arts (located on the UEA campus), and the timber cladding to the west elevation uses reclaimed iroko timber from the original UEA laboratory desks, which were taken out of storage and given a new life by simply cutting and planing them to size.

The passive design principles mean that the building services and distribution systems are far smaller than for a typical building with domestic-sized radiators,

Figure 10.03: The Enterprise Centre at UEA, internal view.

while minimal ductwork lengths use transfer ventilation grilles as part of the heat-recovery strategy.

The result is a lean building, with an overall embodied carbon calculated to be 440 kg/CO_2/m² across the 100-year life cycle. This equates to around a quarter of the lifetime emissions of a conventionally constructed university building of similar size and scale. Comprised of 80% biological materials, the majority of the building can be returned to the biosphere at end of life.

Biohm mycelium insulation

A mycelium is a network of fungal threads that often grows underground. The fruiting bodies of fungi, such as mushrooms, sprout from the mycelium. This thread-like tissue grows fast and tightly together, so it can be coaxed into creating light and strong materials (see Figure 10.04).

Biohm has been developing new strains of fungi that can be used to develop a range of products, including a new building insulation.

Current insulation products are highly carbon-intensive to manufacture and many are petrochemical-based, release volatile organic compounds (VOCs) and are hard to recycle. Incredibly, mycelium is not only entirely biological, but it is naturally fire resistant, has high insulation levels, wicks moisture and sequesters carbon.

Figure 10.04: Biohm mycelium growing in a petri dish.

Figure 10.05 Biohm's mycelium insulation board.

Ehab Sayed, the founder of UK biotech startup Biohm, is now producing a mycelium insulation panel that is the world's first accredited mycelium insulation product (see Figure 10.05).

The researchers at Biohm use a process known as 'directed evolution' to develop new products. The fungi will adapt to stimuli such as differing light or humidity levels, and produce new enzymes. If these adaptations are useful, they can be cloned and developed to create new materials. For example, mycelium naturally contains high concentrations of chitin (usually found in seashells and the exoskeletons of insects), which acts as a natural fire retardant and improves the material's resistance to bio-attack. Mycelium strains can be evolved that produce more chitin to further improve fire resistance.

In keeping with Biohm's circular philosophy, the mycelium is fed on waste from commercial and agricultural by-products that would otherwise be landfilled, incinerated or composted. In the process, it locks in around 70–80% of the carbon from the waste in the material. The 20–30% that is released as carbon dioxide is absorbed by the next batch of mycelium, and is turned into oxygen through photosynthesis. The manufacturing facility has become a carbon sink, as the process actually absorbs and sequesters carbon.

At the end of its service life, the insulation can be fed back into the production process, continually locking the carbon into the material. Biohm is working on take-back schemes with contractors to reclaim and replace insulation, but even if the material is landfilled or composted, the majority of the carbon is stored in the soil. Sayed estimates that every sq. m of mycelium insulation produced sequesters between 0.8 and 1.7 kg of carbon depending on the type of substrate used (different waste streams contain different levels of carbon).

Sayed notes that the mycelium insulation has been tested against construction industry standards by BRE, and is achieving between 40 and 70% lower heat of combustion compared to synthetic insulation such as rigid polyurethane foam or expanded/extruded polystyrene. This means that, in the event of a fire, it would spread 1.5–2.5 times slower than synthetic insulation.

The insulation panels are able to achieve a thermal conductivity of 0.024 W/m²K, which surpasses the values that can be achieved by market-leading materials.

The insulation will start at a price of around £25–30 per m² based on small-scale production, and reduce to below £10–15 as more facilities are opened up in the UK and Europe. Sayed says that even the initial price is comparable to other high-performing alternatives on the market.

This is not only a truly circular product, it is actually regenerative, leaving the planet in a better state than without it. And it actually outperforms the synthetic, linear products in every way.

The next example shows how salvaged biological waste can replace precious new biological materials and shift a material into the technical cycle.

Cross-laminated secondary timber

Cross-laminated secondary timber (CLST) is proving to be an excellent alternative to concrete and steel, and it has far lower embodied carbon impacts as well as sequestering carbon in built structures. However, increasing demand for virgin timber will raise questions about whether sustainable forestry practices can be maintained, or if there is a risk of environmental damage in the long term. There are also supply risks due to increasing competition for land and bio-resources, including fuel and food production.

Meanwhile, demolition and construction in the UK alone produces nearly 1 million tonnes of timber waste a year, and over half of that is solid wood.[5] As noted in Chapter 4, solid timber waste is typically chipped and downcycled into products such as particle board and animal bedding, with limited reclamation of solid timber through salvage yards.

In a circular economy, biological materials should be cascaded through different uses before returning to the biosphere, while avoiding downcycling to lower-grade products (as discussed in Chapter 1). On this basis, a team at University College London has been researching the potential to use salvaged wood to make CLST.

Laboratory testing and modelling of the defects typically found in secondary timber suggested only a small effect on compression and bending stiffness of CLST.[6] Panels can be configured so that timber of significantly lesser quality causes very little reduction in performance. Figure 10.06 shows a prototype CLST panel used as a table top.

As Dr Colin Rose, lead researcher and inventor of CLST says, 'Transforming secondary timber into CLST presents a business case for reusing or upcycling materials on an industrial scale. The process turns low-value materials into a standardised component. It's the kind of product we urgently need to meet the demands of the construction industry, while vastly reducing environmental impacts'.[7]

This is exciting research that shows the potential to save the use of virgin timber, upcycle waste timber, and allow production of CLST locally, rather than relying on those countries with forests.

Even though CLST upcycles waste timber, by gluing it together it moves a biological material into the technical cycle. There are alternative technologies such as friction welding,[8] and biodegradable alternative adhesives that could be used. Then there is an argument that CLST, as a technical material, has a long lifespan and could be designed to be demountable and reused.

Figure 10.06: Prototype CLST panel in use as table top at Chrisp Street Exchange co-working space in Poplar, London.

TECHNICAL MATERIALS

Technical materials are those that are retained within industrial cycles. When used in the structure or fabric of the building, technical components have to be designed so that they can be salvaged for reuse, or re-engineering, with recycling as the last resort. When technical materials are used in plant and equipment, the components have to be designed to be accessible for repair or replacement by 'building in layers' (as discussed in Chapter 6), and can also be designed for upgrade and remanufacture by being modular. In addition, some elements can be leased from manufacturers rather than purchased (see Chapter 12).

The vast array of technical materials that are available to designers, manufacturers and contractors allows the development of incredibly complex components using materials with hundreds of different polymers and alloys. The components are made from materials that are typically not catalogued and can be combined irreversibly with other, very different, materials into what McDonough and Braungart call 'monstrous hybrids'.[9]

This vast array of materials makes recycling difficult. Even metals that are valuable enough to be reclaimed and recycled from buildings can be difficult to recycle at end of life. Allwood and Cullen note in *Sustainable Materials: With Both Eyes Open* that it is not possible to remove the impurities from aluminium, so most recycled aluminium is used as casting alloys rather than wrought aluminium,[10] which has a lower value.

Structural steel sections are robust components that are generally bolted together, making them obvious candidates for disassembly and reuse. Currently, however, they are typically recycled into new steel, rather than reclaimed due to concerns around material quality and a range of practical and logistical challenges. In an attempt to overcome these barriers, the Steel Construction Institute has published a protocol to facilitate structural steel reuse.[11] Cleveland

Steel and Tubes has been pioneering steel reuse, and has a huge inventory of surplus steel tube recovered from steel mills and the oil and gas industry. Many plastics can be segregated for recycling, but only if they are separated from other types. There are two main types of plastics: thermoplastics that can be melted and reformed, and thermosets that cure irreversibly and so cannot be recycled into similar-grade products. Building components use a wide range of plastics, including thermoset plastics such as polyisocyanurate (PIR) and phenolic foam for components such as insulation. These materials are typically sent to landfill and incinerated rather than being recycled.[12] Thermoset plastics can be downcycled into a lower-grade product by crumbing and remoulding with new material.

Some manufacturers have recognised the problem with a wide range of materials and have worked to reduce them: Hewlett-Packard has been steadily reducing the number of plastics used across its product range from 200 to six to make it easier to recycle them at end of life.[13]

Biomimicry and 3D printing

According to Pawlyn in his book *Biomimicry in Architecture*, 'Nature uses a very limited subset of the periodic table whereas we use virtually every element in existence, including some that would be better left in the laboratory.'[14] Janine Benyus (author of *Biomimicry: Innovation Inspired by Nature*) states that: 'life builds from around five commonly used polymers'. From these polymers, nature has created a myriad of complexity, from abalone shells that form a material stronger than the toughest ceramic,[15] through to materials such as cotton, valued for its softness and flexibility.

Nature uses structure and form to create stiffness, flexibility and even colour from its limited palette of materials. Structural elements such as bones and tree trunks use complex forms to create light, efficient structures from microstructures that start at the molecular level. A butterfly's wing is coloured not by pigmentation but by a microstructure that refracts light.

Additive manufacture, or 3D printing, allows components to be built up in layers using industrial robot technology created from CAD files or 3D scanners. The technology uses a range of materials including plastics, ceramics and metals to create forms with complex internal geometry. Designers and contractors are now experimenting with 3D printing of building components using traditional construction materials such as concrete, aluminium and glass.

The technology represents both an opportunity and a threat to the circular economy. The main threat is that there is the potential to create a highly complex mix of materials that would be impossible to separate at end of life. The idea of printing houses from cement blended with recycled rubble, fibreglass, steel and binders, as demonstrated by a construction firm in China, leaves the legacy of a solid chunk of composite material that will have to be demolished and crushed at some point.

The opportunities, on the other hand, are exciting. Additive manufacture means that components can be shaped and formed through addition, rather than the traditional industrial approach that involved removing material to create the desired design. Additive manufacture avoids the 'yield losses' associated with off-cuts, machining, components rejected due to defects and so on. Research by the University of Cambridge shows that the global yield losses from turning liquid metal into a fabricated product are 26% for steel and 41% for aluminium, so there is the potential for considerable savings.[16] This could be combined with the idea that products can be made with the exact amount of material required to perform the function required, so structures could have material added at points of increased stress and even designed with the complex internal geometries demonstrated in nature.

The potential applications of 3D printing inspired by biomimicry are explored in Michael Pawlyn's book. This includes the idea inspired by the butterfly wings mentioned above: creating a nanostructure from glass that performs a similar function to the low-emissivity coatings for high-performance facades. This would mean the glass element would be made from only one material, making it far easier to recycle at end of life. Applying this approach of using structure to create different performance characteristics of materials, rather than using more polymers and more ingredients in materials, should contribute towards a more circular economy.

K-Briq

Here is a challenge: create a brick that uses recycled demolition and construction waste and has one-tenth of the embodied carbon of a conventional masonry unit, while still being structurally sound. Kenoteq has risen to the challenge by creating the K-Briq, an unfired brick made from demolition and construction waste and a secret binding agent developed at Heriot-Watt University. Unfired bricks have a fraction of the embodied carbon impacts of kiln-fired bricks. They are also hydroscopic, which helps to regulate internal humidity, and they come in a wide range of colours, as shown in Figure 10.07.

Traditional earth construction methods were the inspiration for the development of the K-Briq. However, rather than using old techniques that exploited raw virgin materials, Kenoteq targeted a circular approach where resource efficiency was achieved by reusing demolition and construction waste as the main input material (over 90%) of the K-Briq.

K-Briq will address three key problems: the rising demand for bricks for new homes, the need to slash embodied carbon, and what to do with all that demolition waste.

Figure 10.07: K-Briq: unfired bricks come in a range of colours.

Finishes and fixings

Finishes can inhibit both the maintenance and upgrade of buildings, and the deconstruction and reclamation of materials or components. Carpets represent a significant environmental impact as they are frequently replaced long before the end of their service life, and a high proportion of the waste is not recycled.[17] Using carpet tiles in commercial buildings is common practice and allows layouts to be changed, worn-out tiles to be replaced and services under raised floors to be accessed.

Carpets are a good example of how the principles of designing for adaptability and deconstruction can be applied, by removing the need to glue them down. Carpet manufacturers have developed systems to eliminate the need for adhesives – for example by using high-friction lightweight coatings and adhesive-free corner pads to bind the tiles together.[18] These systems allow carpet tiles to be lifted easily for access, replacement or removal at end of life. Similarly, there are ceramic floor tiling systems that are laid on to a rubberised grid system. The system avoids any need for grout or adhesive, does not release fumes and can be installed over an existing floor. It also allows cracked tiles to be easily replaced, and they can be reclaimed at end of life.[19]

Using materials that have an 'inherent finish' helps to maintain the purity of the material. Timber such as oak, larch or western red cedar can be used untreated for external cladding, as it weathers naturally. Aluminium, stainless steel or Corten steel are inherently resistant to weathering.[20] There are timber treatments that extend the life of the wood and allow it to be used externally while not using toxic chemicals or contaminating the material, such as the Thermowood heat treatment and Accoya. As Pawlyn explains in *Biomimicry in Architecture*, both techniques make the wood indigestible to microbes. The Thermowood treatment exposes the timber to intense heat, and Accoya uses a process of acetylation using a benign chemical (acetic acid).[21] For internal finishes, using inherent or natural finishes that do not offgas air pollutants such as VOCs benefits both the internal environment and avoids pollution of finishes during manufacture, application and at end of life.

CERTIFICATION SYSTEMS

Selecting materials and products that are compatible with the principles of a circular economy is a complex and involved task. The shortcut is to use certification schemes that analyse and assess products against these criteria and label products according to their relative environmental performance.

There are many certification schemes that consider different aspects of the impacts of materials and products. Some consider the carbon emissions associated with products, while others focus on products with low chemical emissions that may pollute the internal environment.

ISO 14025 sets standards for voluntary environmental labels and identifies three types of labels, ranging from third-party labels that consider the whole life cycle through to self-certified declarations of performance.

Examples of labels that align closely with the circular economy principles include the Cradle to Cradle® (C2C) Certified™ Product Standard and Natureplus®. Natureplus® assesses products against specific criteria for each product group. The criteria are focused on: 'The protection of limited resources by the minimisation of the use of petrochemical substances, sustainable raw material extraction/harvesting, resource-efficient production methods and the longevity of the products. Therefore, building products made from renewable raw materials, raw materials which are unlimited in their availability or from secondary raw materials will be favoured for certification.'[22] Natureplus requires the manufacturer to make a declaration of all the input substances and proof of origin of all the input materials. It includes a list of banned substances that are damaging to health or the environment. It also sets out rules for deconstruction and recycling at end of life, very similar to those set out in Chapter 9.

Cradle to Cradle® (C2C) CertifiedTM Product Standard

The Cradle to Cradle® (C2C) Certified™ Product Standard was originally developed by McDonough and Braungart, and aims to implement the cradle-to-cradle philosophy (see Chapter 1). The scheme assesses products under the following categories:[23]

- Material health: knowing the chemical ingredients of every material in a product, and optimising towards safer materials.
- Material reutilisation: designing products made with materials that come from and can safely return to nature or industry.
- Renewable energy and carbon management: envisioning a future in which all manufacturing is powered by 100% clean renewable energy.
- Water stewardship: managing clean water as a precious resource and an essential human right.
- Social fairness: designing operations to honour all people and natural systems affected by the creation, use, disposal or reuse of a product.

Under the standard, products are rated Basic, Bronze, Silver, Gold and Platinum, showing the level of achievement against each of the above categories. This provides manufacturers with the opportunity to achieve an entry-level certification and a pathway to the highest rating.

Cradle to Cradle® certification includes a list of chemicals that are banned due to their tendency to accumulate in the biosphere and lead to irreversible negative human health effects. These are split into technical and biological nutrients. The separate lists allow for the use of some substances, such as lead or cadmium, in materials where it is unlikely that there will be exposure to humans or the environment. For example, lead is used in cast aluminium, but it does not migrate out of the material. However, lead should not be used in biological nutrients, to ensure that it is not released into the biosphere.[24]

NASA Sustainability Base

The NASA Sustainability Base in California (completed in 2012) was designed by William McDonough + Partners with AECOM as the architect of record and engineer of record (see Figure 10.08).

A rigorous selection process was applied to the choices of materials and products that were used on the building to implement the cradle-to-cradle vision for the project. Cradle to Cradle® Certified™ products were used when available, and when they were deemed cost-effective. The other products were evaluated by McDonough Braungart Design Chemistry (MBDC) using the Cradle to Cradle® criteria. Figure 10.09 shows the C2C products that were selected, including elements of the cladding, the hardtops and the office chairs.

The main building elements are formed from steel, glass and aluminium, selected for their high recycled content and their ability to be reclaimed or recycled. The lobby areas reuse oak flooring from a transonic wind tunnel on the NASA Ames Campus.[25] The steel structure is designed for deconstruction (see Figure 10.10) and the exterior cladding consists of prefabricated modular units.

The Cradle to Cradle® (C2C) Certified™ Product Standard is gaining traction in the construction industry, with recognition in two environmental assessment

Figure 10.08: NASA Sustainability Base.

methods: RICS SKA Rating and LEED v4. As it gains in momentum, the number of construction products that are certified is increasing, which provides more opportunities for designers to specify compliant products.

Whatever eco-labelling system is used to help with product selection, the main consideration should be to check that it is third-party verified and that it considers the whole life of the material or product, including the manufacturing and in-use impacts, and the ability to disassemble, reuse or recycle at end of life.

CONCLUSION

Materials and component selection form the bedrock of a circular economy. Materials should be selected based on the following:

- Decide whether the element will have a long or short lifespan – in reality, not according to a table – and select the appropriate material that matches that lifespan.
- Determine whether the materials are technical or biological. This decision depends on the lifespan and the likely fate of the element as well as its function.

- Retain the purity of the materials by avoiding mixing biological and technical substances, and ensure that biological materials are not contaminated by toxins that inhibit them being returned to the biosphere.
- Substitute technical materials for biological materials, particularly where the component has a short life, or where technical materials could be difficult to manage at end of life.
- Select reused and reclaimed components and materials.
- Ensure that elements can be disassembled at end of life with the ability to upgrade or remanufacture or reclaim the materials.

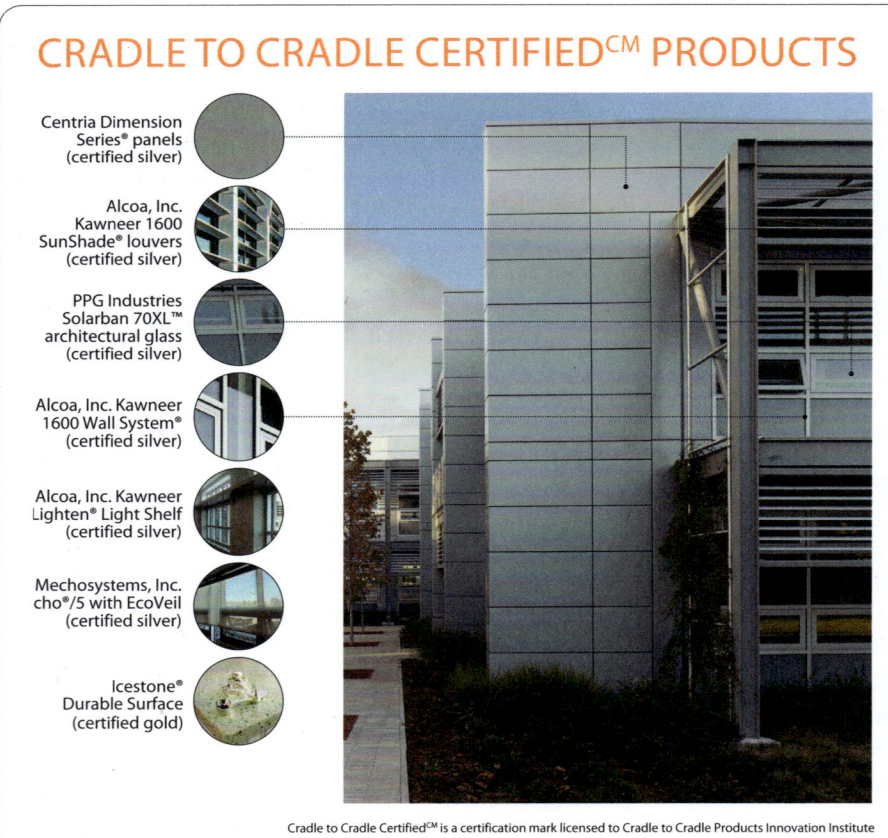

CRADLE TO CRADLE CERTIFIED^{CM} PRODUCTS

Centria Dimension Series® panels (certified silver)

Alcoa, Inc. Kawneer 1600 SunShade® louvers (certified silver)

PPG Industries Solarban 70XL™ architectural glass (certified silver)

Alcoa, Inc. Kawneer 1600 Wall System® (certified silver)

Alcoa, Inc. Kawneer Lighten® Light Shelf (certified silver)

Mechosystems, Inc. cho®/5 with EcoVeil (certified silver)

Icestone® Durable Surface (certified gold)

Cradle to Cradle Certified^{CM} is a certification mark licensed to Cradle to Cradle Products Innovation Institute

©William McDonough + Partners. All rights reserved.

Figure 10.09: NASA Sustainability Base, Cradle to Cradle® Certified™ products.

These decisions are complicated, and require a focus on procurement and a lot of information about the substances in materials and components. Labelling and rating schemes can reduce many of the complexities in these choices by vetting and rating components for designers. This level of scrutiny of materials makes it essential to build partnerships and collaborative relationships with the supply chain. This is explored further in the case study in Chapter 13 (Park 20|20).

By implementing these ideas, buildings can be turned into materials banks that will enable future generations to salvage valuable materials. Equally, existing waste streams can be turned into valuable materials that can be salvaged and used in creating new designs. Turning waste into a resource is explained in the next chapter.

Figure 10.10: NASA Sustainability Base, showing steel 'exoskeleton' structure.

TURNING WASTE INTO A RESOURCE

The goods of today are the resources of tomorrow, at yesterday's resource prices.

— Walter R. Stahel

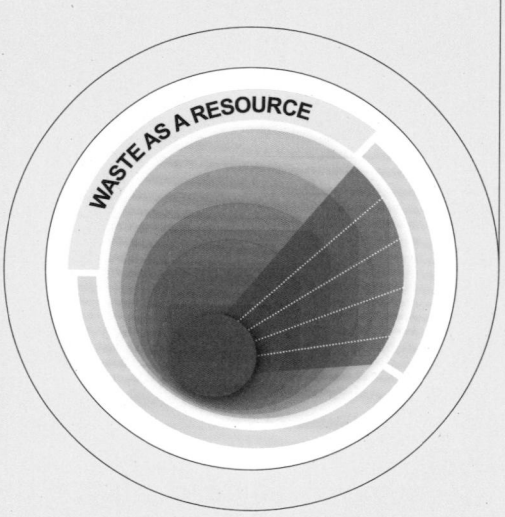

WASTE AS A RESOURCE

In a circular economy, waste becomes a resource. Current demolition practices are driven by the pressure to tear down old buildings quickly and safely, and the lack of demand for reused components means that the majority of materials are downcycled. By designing buildings for disassembly and deconstruction, the residual value of an obsolete building can become positive, and there is more incentive to reclaim components and materials.

Redefining waste as a resource requires systemic changes to the industry. To start with, building design has to enable materials to be reclaimed by designing for deconstruction/disassembly, and materials in the building have to be catalogued so that the residual value can be calculated and new markets for components can be established. When the building is due to be demolished, more time, space and labour have to be allowed for deconstruction, and new techniques are needed to recover valuable components and materials from existing buildings. Lastly, there needs to be an active market for salvaged components and materials that covers the costs of disassembly, storage and resale. Creating a market for salvaged stock means that reused components have to be as readily available, attractive and fully certified as new products. And there has to be a change of mindset in designers to tailor their designs to incorporate reclaimed materials and components in their new designs.

This may seem impossible, but other industries are taking steps down this path, and the case studies in this book show how some building projects are starting to think differently by using reclaimed products in building design and fit-outs, and by designing for deconstruction.

To support and enable this shift in thinking, new businesses have emerged that rely on 'urban mining' and the deconstruction of buildings.

Urban mining is a vision for the future of the construction industry. Our towns and cities contain highly processed, valuable resources from all around the world that have been collected and stored in highly concentrated quantities. Currently, these precious resources are mostly bound together irreversibly and are being crushed and shredded and sent out to surrounding areas, or even abroad, for processing and downcycling.

Urban mining recognises the opportunity for cycling those materials within the urban environment, saving raw materials and avoiding waste. There is a huge stock of buildings, elements and materials that should be exploited before new raw materials are required. When viewed at the city scale, instead of at the project level, reclamation of resources becomes more practical because buildings can become the materials banks in which to store components and materials (see Figure 11.01).

Successful urban mining requires data on the locations, quantities and potential release dates of products and materials, new techniques to extract them, new business models to match supply and demand, and brokers to remanufacture and warranty products.

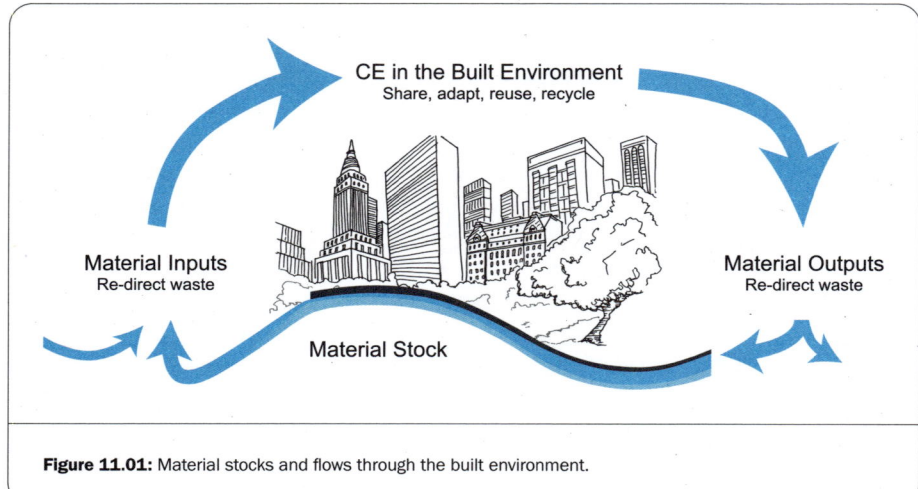

Figure 11.01: Material stocks and flows through the built environment.

For new buildings, materials passports are being used to start to catalogue the raw materials in each element.

RAW MATERIALS PASSPORTS

Thomas Rau is an architect based in Amsterdam. As well as being an advocate of service models over ownership (see Chapter 12), Rau proposes that all the components of a building should have a 'raw materials passport' to give waste an identity. As he says: 'Waste is a raw material without identity.' The idea behind the name is that a 'passport gives you identification – you can't leave the country without it'.[1] If materials do not have a description, then they are likely to be downcycled and much of the value will be lost.

Building information modelling (BIM) enables constructors to create detailed information about the components of a building and communicate that to operators and those responsible for refurbishing or dismantling the building.

For the existing building stock, being able to catalogue and identify materials is a key step towards effective urban mining. Researchers are using digital techniques to catalogue the resources stored in existing buildings, and have started with looking at a modular component that is used extensively in our cities and towns – bricks.

HOW MANY BRICKS ARE THERE IN BRADFORD?

A study by Krausmann, et al.[2] estimates that demolition of bricks, stones and tiles alone created 1.3 billion tonnes of construction and demolition waste (CDW) globally, while 3.2 billion tonnes were added to the built environment in 2010.

A city-wide inventory of the resources tied up in the building stock would help supply chain stakeholders to identify and unlock the potential value of the materials and the structural products when buildings are at the end of their service life, instead of carrying out destructive demolition. A research project has mapped the brick resources of the city of Bradford and identified the different types of bricks, their locations and their embodied carbon. This will provide prospective urban miners with data on the stocks of materials and products available on a site that is earmarked for demolition.

In the REBUILD project,[3] all the different types of bricks were identified ranging from traditional hand-made bricks through to contemporary, modern bricks as shown in Figure 11.02.

This spatiotemporal map provides the locations of the potential stocks and supply of reclaimable bricks, including embodied carbon, building typologies and a crucial time-bound layer to assess potential future end of life and release rates.

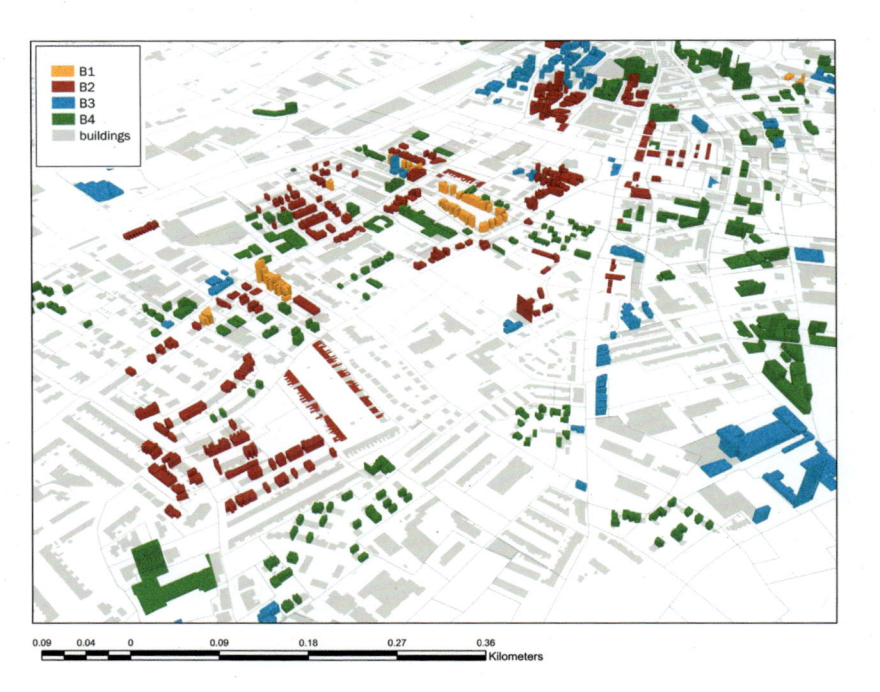

Figure 11.02: This shows the density map of four brick types (B1: hand-made, pre-1850; B2: mechanised, 1851–1945; B3: imperial BS, 1946–1970; and B4: modern, 1971–present).

So, how many bricks are there in Bradford? According to the study, there are an estimated 209 million.

It is one thing to know where the bricks are and roughly when they may be available, but there is still the thorny question of how they can be separated from the cement for reclamation. To overcome this, urban miners need new techniques to address these legacy methods of construction.

PUNCHING OUT BRICKS

Bricks laid in modern ordinary Portland cement (OPC) mortar have been considered too difficult to reclaim, as it is labour-intensive and time-consuming, and there is a risk of damaging the bricks. However, the REBUILD project team has found a potential solution. Research led by Professor Wang and Dr Chen at Manchester University has created novel techniques and processes inspired by a hole-punching technique used in manufacturing goods.

In their laboratory, they devised a special steel template to punch the mortar out of the bricks using an Instron loading machine. The reclaimed bricks are not damaged and are suitable for reuse, with almost the same compressive strength as new bricks[4] (see Figure 11.03). Although no commercial machine exists yet, the punching technique can be readily automated and turned into a site-based technology, which would open up the potential to reclaim a proportion of the 2.5 billion bricks that are crushed every year in the UK alone.[5]

It is not just bricks that are crushed and downcycled. Concrete structures suffered the same fate, until the invention of the Smart Liberator.

MAKING NEW CONCRETE FROM OLD

Concrete forms are, quite literally, the foundations of the modern world. Concrete is intrinsic to the way we live, providing the structures for virtually all our buildings, and providing the infrastructure that supports our way of life.

Concrete is the most widely used construction material in the world, and the second most used material on the planet, after water.[6] Every year over 4 billion tonnes of cement are produced, and concrete is responsible for around 8% of all global carbon emissions.[7] As well as the emissions, sourcing the raw materials requires global mining and dredging operations, and huge volumes of water, often extracted in water-stressed areas.[8] To exacerbate the problem, there is now a global shortage of the right type of sand required for construction, typically found in riverbeds and beaches.[9] Sand from deserts is plentiful, but it is too smooth and rounded to bind well in concrete.

60% of the carbon emissions from cement production are caused by a chemical reaction when limestone is converted to clinker,[10] which is then processed into cement. These emissions are inherent to the production of cement.

The concrete industry is well aware of its impact, and in the UK has already delivered impressive reductions in absolute carbon emissions from concrete

Figure 11.03: Punched-out bricks, successfully reclaimed.

production and has set out 'A roadmap to beyond net zero'[11] that includes carbon capture, usage and storage technology, and innovative concrete mix design to meet the UK Government's net zero target.

Concrete in existing structures is typically downcycled by being crushed and used as fill for road bases and pile mats; it is not turned back into structural concrete. A new solution is solving this problem and creating circular concrete for the first time.

Circular concrete: the Smart Liberator

Concrete has been embedded in three generations of the Rutte family business from the time that Joop Rutte started transporting construction materials around Amsterdam after the Second World War. His grandson Rene Rutte's company was the first Dutch company to use mobile electric crushers on demolition sites to create gravel and sand from waste materials. However, the local concrete manufacturers did not want to use the crushed materials to make their concrete, preferring to use virgin materials. So Rene set up his own concrete plant to use the materials himself, creating precast concrete boxes to house underground recycling storage facilities.

Sven Hiskemuller van der Zijden, the sustainability advisor for Rutte Groep, and Rick Rutte, the latest generation of the Rutte family, explain how their desire to be a circular and sustainable business inspired the idea of circular concrete.

According to Hiskemuller van der Zijden, the cost of concrete doubled in 2019 in the Netherlands because of climate change. The country's main source of gravel and sand is from dredging the Rhine, and the lack of rainfall in 2019 meant the river was too shallow to allow the dredgers to operate.

The team realised that using recycling aggregates and cement replacements (such as GGBS and PFA) was only going to scratch the surface of the environmental impacts of concrete, and that a more innovative solution was required.

The breakthrough was the realisation that there is unhydrated cement in existing concrete. This is the cement that has not reacted with the water and so is still raw cement. Dutch innovators Koos Schenk and Alef Schippers together with the Rutte Groep realised that, if they could crush old concrete to make the ingredients for new concrete, including the gravel, sand and unhydrated cement, then they could solve both the problem of what to do with the old concrete and avoid having to mine raw materials and manufacture new cement.

One of the major obstacles was that the concrete from demolition was often contaminated with bricks, wood, glass and many other materials. The innovators experimented for two years with various crushing techniques. In search of a solution, they looked to the very industry that they were trying to replace. The mining industry has had to contend with similar problems to separate unwanted elements from the desired material (e.g. brown coal from coal).

A final hurdle was how to separate the unhydrated cement from the hydrated cement. Without this, the vision of creating new structural concrete from old could not be realised. Through research and experimentation they discovered that the cement that had reacted with water to form concrete was lighter, and that an air separation technique could be used to siphon off this element.

They drew on these techniques and created the Smart Liberator, a crusher that is able to liberate gravel, sand, hydrated and unhydrated cement from demolition waste. The Smart Liberators are now in use, and are processing 422,000 tonnes per year of demolition waste that has been mined from the city (see Figure 11.04). Incredibly, the break-even point is after only six weeks of operation, after which the initial investment in the equipment pays for itself by avoiding the cost of the raw materials and the cost of the disposal of waste.

The team estimate that the current process reduces the carbon emissions associated with concrete production by over 90% compared to conventional concrete. This is as well as avoiding the need to dredge and mine new raw materials, and solving the problem of how to treat low-grade demolition waste. As a bonus, the huge industrial units are covered with photovoltaic panels, supplying electricity for the whole operation, and the roofs are used to collect rainwater, providing water for the precast concrete made on site.

Next, the team are doing research with Leuven University to develop a microwave technique to reactivate the hydrated cement. If this is a success, it would be the final piece in the puzzle towards reusing all of the concrete waste to create new concrete.

The Smart Liberator is an excellent example of how a seemingly intractable problem can be solved, and is already turning a material that is locked into a

Figure 11.04: The Smart Liberator.

downcycling spiral into a truly circular product that can be renewed indefinitely. In the words of Hiskemuller van der Zijden: 'mining has become urban mining'.

DECONSTRUCTING BUILDINGS

Buildings are typically not deconstructed for lots of good reasons: it takes longer, it requires more labour, it costs more money, there is nowhere to store the reclaimed materials, and it is harder than building new. However, there are organisations across the world that are not only doing it but are competing with conventional demolition companies and disrupting the market in the process. It is a rapidly growing industry that is providing new jobs, supporting communities and creating profitable businesses, as well as providing huge environmental benefits by turning waste into a resource.

These new businesses are creating two revenue streams: they get paid to deconstruct a building and then get paid again to sell the components they harvest.

One example of these disruptive organisations is the Building Deconstruction Institute, in Washington.

The Building Deconstruction Institute

Dave Bennink has been deconstructing buildings for over twenty years, and has helped people in 44 States and four Canadian Provinces to set up reuse companies, create jobs and deconstruct buildings. Bennink set up the Building Deconstruction Institute in 2018 to upskill the industry and share his experience of deconstructing over a thousand buildings, from blighted inner-city homes to large warehouses.

The first building he deconstructed took three and a half weeks, now it takes three and a half days. Considering that it would have taken a day to demolish the house and a day to clear the site, it is not that much slower. Bennink's approach is what he calls 'hybrid deconstruction', combining the best of demolition strategies and the best of deconstruction strategies.

The first ingenious step is his departure from conventional deconstruction. Taking stuff apart requires a lot of labour to remove nails and other fixings, and all there is to show for the effort is some usable material and some debris. Instead, Bennink's team finds someone who wants to construct a new building and matches them up with a building to be deconstructed. He then cuts sections from the existing building of the size required to make the new building. He calls these panels 'post-fabs' (as opposed to prefab). The section may have a window cut into the wall, and it may have spray-foam insulation, which is difficult to separate and recycle. By reusing the whole element, the issues around disassembly are bypassed. The BDI has provided the post-fab elements for over a hundred buildings, all created from existing buildings that were deconstructed. This can work for anything from engineered trusses, through to whole staircases.

The assemblies that are reclaimed are used for buildings such as storage sheds, and provide a much higher standard than anything that would be available from a store.

The second ingenious step is to realise the value of the components and materials: BDI is paid to disassemble the building, and then paid again to sell the components and materials. When Bennink's team bid to deconstruct a building, they bid 'low' because they want to stay competitive and break even, and then get the building materials to sell. They focus on volume and the mainstream market. Some reclamation companies charge a premium to cover the cost of storing and processing high-value materials, but BDI markets affordable building materials such as 2 x 4 in. timber wall studs that they sell at a dollar each. Of course, if they do find desirable building materials with a beautiful patina, then they will charge a premium for those items.

These two elegant steps have provided a resilient business model that helps the environment and the economy and addresses social inequality.

The formula has proved so effective that Bennink has helped to set up many other businesses across the United States. This business model has provided multiple simultaneous benefits relating to training, employment, the environment and provision of affordable materials. When Covid-19 hit, the organisations became incredibly busy because they were selling affordable materials.

The idea of cutting customised chunks out of buildings for reuse is also the subject of research in Kerkrade, a town in the Netherlands.

The Super Circular Estate, Kerkrade
The EU Urban Innovative Actions (UIA) project, the Super Circular Estate, in Kerkrade, has gone beyond simply mining materials from existing buildings, by cutting customised section from an existing 10-storey block of flats built in the 1960s and using them as a structure for new housing (see Figure 11.05).

The consortium, led by the Municipality of Kerkrade, calculated that a conventional demolition process would result in 1,380,000 tonnes of demolition waste that would be downcycled into secondary aggregates for road bases and backfill, losing 287,000 tonnes of embodied carbon.[12]

The consortium deconstructed the tower block building and cut out tunnel-shaped concrete sections that were used to form the structure of new houses. The lightweight concrete partition walls and doors were reclaimed, concrete chunks and reclaimed bricks were used in the facades, and additional structural walls used recycled aggregate.

As noted by Elma Durmisevic, the UIA expert on the project, new deconstruction strategies can radically increase the reuse of building elements, but this only provides one more use for the elements – a 'stay of execution' before downcycling. To create truly circular buildings, new buildings will have to be designed for disassembly. This inspired one of the consortium partners

Figure 11.05: The Super Circular Estate, Kerkrade showing three circular houses and the concrete tower block in the background.

(Dusseldorp – which deconstructed the existing building) to propose and prototype a reversible concrete block that interlocks and stacks with no need for mortar.

The contractor responsible for the construction of the test buildings, Jongen Bouwpartners, proposed a prefabricated concrete system with demountable connections that is being used for the construction of fifteen new circular homes in Kerkrade.

The melding of demolition and new-build contractors in a research project has yielded innovations that can be tested away from the daily pressures of a live construction project, and has also provoked new discussions within the industry.

The next question is how to combine the two industries and create ways for new buildings to incorporate existing resources and materials.

CREATING A MARKETPLACE FOR SALVAGED MATERIALS

A project that was developed during the UK Green Building Council (UKGBC) Future Leaders programme[13] proposed an online hub to create a marketplace for building materials. The hub would document the materials used in the building and make the information accessible to interested parties. The materials would be documented in a materials passport which would contain a quantified list of construction materials and modular elements (e.g. windows) along with their recovery rating.

The proposed hub would allow users to place bids on materials and see a date when they might be available. It could help to assign a value to materials

held within existing buildings, with parties being able to purchase an interest in the material in advance of it being salvaged from the building.

The UKGBC Innovation team argues that building owners often know that a building is to be demolished well in advance, but the actual demolition time is usually very constrained, allowing little or no time to find a market for salvaged materials. Selling the materials in advance could incentivise the building owners to allow more time to deconstruct the building and ensure that the components are salvaged intact. In the long term, the hub could also help to encourage the design of buildings that are simple to deconstruct, enabling valuable components and materials to be reclaimed at end of life. Creating a guaranteed market for reclaimed materials prior to demolition would also provide an economic incentive for demolition contractors to salvage components and materials.

Some entrepreneurial organisations have created their own platforms to share resources.

Globechain

May Al-Karooni was working for a large bank in London when her inspiration struck. The bank was moving offices literally across the road, and when the facilities team asked Al-Karooni and her team to pick out new furniture, new ICT and finishes, she asked what they were doing with all the existing stuff and why could it not be reused, or at least be sent to charity. But the facilities team did not have the time or the resources to find new homes for all the items.

Al-Karooni realised that there was a pressing need to digitise the waste industry, and an opportunity to divert vast amounts of unneeded items away from being downcycled or sent to landfill.

Al-Karooni went on to become the founder and CEO of Globechain, a reuse marketplace that connects corporates to charities, SMEs and people to redistribute unneeded items to those that need them, generating social impact data for its members.

Having worked in the corporate world, Al-Karooni understands what makes companies tick: the charities that are signed up on the network all have the correct procedures in place to ensure that they can collect the items and provide all the requisite paperwork; and the masterstroke is that the marketplace generates a social impact (environmental, social and governance, or ESG) report that is sent back to the corporate. The reports include the kilos diverted from landfill, who took the items, what they were used for, and how they were used. For example, it could show how many apprentices have been helped by the donation of an over-order of bricks through reuse, or how the emergency lights that were going to be crushed and incinerated have, instead, been used as lights for emergency hospitals in a war-affected part of the world. This is invaluable data for corporates when they are reporting their social impacts in their annual reports.

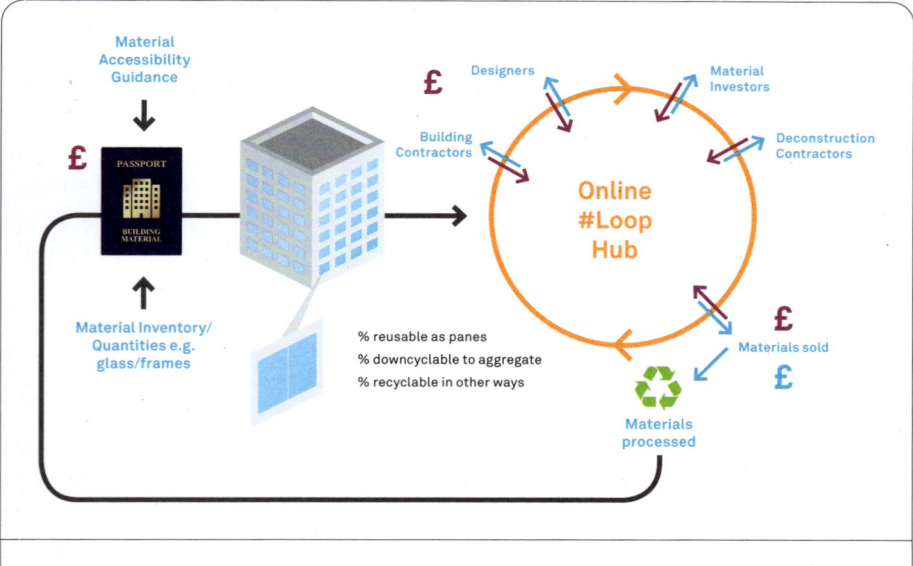

Figure 11.06: An online marketplace for materials, proposed by UKGBC.

Globechain is now operating in the UK, Spain, the UAE and New York. It has over 10,000 members and has diverted over 6.1 million kilograms of resources from landfill, creating savings of over €4.4 million for charities.

BROKERS FOR A CIRCULAR ECONOMY

More than anything, the circular economy needs brokers to bridge the gap between potential resources that can be reclaimed, and the designers and contractors who want to use them.

Brokers break through all the barriers to reclaiming components and can even offer cheaper, better-quality products than new items.

Paint360 – remanufactured paint

An estimated 55 million litres of waste decorative paint is generated in the UK each year. The vast majority of it is disposed of through incineration, at a considerable cost to local authorities. It seems illogical to be not only wasting perfectly good paint, but also having to pay an estimated £20.6 million to dispose of it.

Paint360 is a social enterprise that takes unwanted and waste paint and re-engineers it into a new product. It takes the paint directly from waste disposal companies, blends it and disinfects it (once the tin is opened it is susceptible to bacterial infection). This provides the raw material that can then be re-engineered into different products, depending on both the properties of the paint and the market demand. The colour and performance may be changed, but the raw mixture is turned into new decorative paints or even specialist products to treat mould.

This all sounds great, but is the re-engineered paint really as good as new? Because most people buy premium brands of paint, this is what ends up in the waste stream. So, re-engineered paint is based on a high-quality raw ingredient and it therefore competes on quality.

Paint360 is already distributing through three large builders' merchants and directly to leading UK construction firms. It has also developed a retail brand for the domestic market called Paintsavers.co.uk that uses unwanted paint as well as waste paint. This includes 'off-specification' or damaged goods from manufacturers.

Established in 2013, the organisation has already helped more than 30 young people with barriers to employment into full-time work, and is now looking to expand into other regions of the UK that have high unemployment. There is so much waste paint that it has to turn some of it away. In response, the founders are planning to quadruple production on their new Halesowen site up to 2 million litres per annum, and are creating a new site on an eco park in South Wales.

As Lee Cole, managing director and co-founder says, their 'paint has to compete on quality and price compared to premium brands so clients get the social and environmental stuff for free'.

Diverting a hard-to-manage waste from incineration is good news, using it in place of raw materials to create a new product is better, and helping young people into employment is inspired.

Remanufactured raised floor tiles – RMF Eco-Range/RMF E-Coated™

RMF Installations and Services Ltd (RMF) is an exemplar broker within the circular economy, which is capitalising on one of the few truly standard, modular products currently used in buildings. The raised floor tile is a standard-sized product that is ubiquitous across the commercial office sector in the UK. Raised floors are inherently demountable to allow access to the floor void, making them a perfect candidate for reclamation.

Raised floor tiles are regularly stripped out of buildings and often end up in landfill (see Figure 11.07). They are hard to recycle as they are made from a composite of particle board wrapped in galvanised steel. The timber makes them heavy, they often pass through the jaws of the shredding machine, and scrap yards frequently charge for processing them as it is difficult to extract the value of the steel.[14]

RMF has developed a raised floor reclamation system over the past seven years. It will collect tiles from demolition and construction sites, clean them up, test their integrity, rewarranty them, and sell them for substantially less than the equivalent new product.

Figure 11.07: Remanufactured raised floor tiles being laid.

Providing a warranty for the tiles is critical to them being procured, and sets them apart from many other reclaimed products. The tiles are randomly point-load tested over 24 hours to ensure their integrity, enabling RMF to provide a 25-year warranty.

As Simon Middleton, co-owner at RMF, notes, 'the raised floor tiles manufactured in the late 1980s and 90s generally still outperform new tiles manufactured today'. However, there was a period when carpets where glued down by pouring adhesive across the raised floors. Sometimes the glue even dripped between the tiles, meaning that they cannot be lifted. When raised floor tiles are received at RMF to recondition, the adhesive is hard to remove and means that the end finish after removal of the adhesive is matt.

According to Simon, the matt finish is one of the biggest barriers to their reuse. So, although the product is cheaper, has virtually zero embodied carbon compared to new tiles, and is tested and warrantied, the new tiles are preferred because the completed Cat A offices have a shiny raised floor that glints in the sunlight and helps to attract tenants. This ignores the fact that the floor will be covered up with carpets and other finishes before anyone moves in.

To break down this final barrier to reuse, RMF has launched its RMF E-Coated™ system, utilising a zero-VOC coating that gives the reclaimed panels a consistent colour and finish.

The tiles come with an independent Environmental Product Declaration that demonstrates the life cycle benefits of the product compared to new systems, and will ensure that they are recognised by environmental assessment schemes such as BREEAM.

RMF holds a stock of around 100,000 reclaimed tiles, and the stock is always replenished as the rate of office strip-outs matches the demand for new tiles.

A product that diverts waste from downcycling or landfill is cheaper, has a fraction of the embodied carbon, can outperform new tiles, is fully warrantied, and can be reused again and again. RMF therefore meets the gold standard as a circular economy broker.

CONCLUSION

Turning waste into a resource means stopping it from becoming waste in the first place. This means that materials have to be clearly identified, catalogued and separable to allow a value to be attached to them. Designers have to think about using reclaimed materials and how they can be salvaged at end of life, contractors have to be able to source materials and components that are comparable to new products, and demolition contractors and manufacturers have to become brokers who salvage and sell resources. This can only work if new, vibrant networks are set up to make links between different industries and create a flow of materials that keeps resources in circulation. The circular economy needs a hub around which to turn.

12

CIRCULAR BUSINESS MODELS

In order to change an existing paradigm you do not struggle to try and change the problematic model. You create a new model and make the old one obsolete.

— Buckminster Fuller

CIRCULAR BUSINESS MODELS

The linear economy is based on manufacturing products and selling them to a consumer, who then consumes them and is responsible for disposing of them at end of life. The circular economy needs new business models to retain the value of products and help to close the loop.

The business models range from full performance-based models through to take-back and remanufacture. In the performance-based models, the manufacturers retain ownership of the products while providing a service to the customer. The take-back model proposes selling products that are designed to be returned and remanufactured with incentives such as deposits, or guaranteed buy-back, to encourage customers to return the used products when they are no longer required.

PERFORMANCE, NOT PRODUCTS

The industrial economy is transforming from a production-based model into a more intelligent performance-based model. Yet despite the proven benefits that selling performance provides, too many managers and policy makers still focus on designing, manufacturing, and selling goods using costly economic models and production methods.

Walter R. Stahel[1]

Walter R. Stahel has long promoted the concept of selling performance rather than products, so the consumer becomes a user and ownership is replaced by stewardship. There is an increasing shift towards selling performance over purchasing products, with people leasing printers, cars, office space and even clothes (e.g. Mud Jeans). Selling performance means that the manufacturer retains ownership of the materials, therefore securing their supply of components and materials in the future. This helps to maintain the value of the components, as manufacturers are more likely to:

- design their durable products to have a long service life, with the ability to repair and upgrade, or even remanufacture them
- design their consumable products to be simple to disassemble and recycle, or to return to the biosphere
- ensure that toxic materials are easy to separate and reclaim for reuse, or are designed out of the products.

These arrangements can be good for businesses, as they create long-term relationships between manufacturers and customers, rather than one-off interactions.

RAU Architects and Turntoo

Thomas Rau's design philosophy rests on the premise that the planet is a closed system, and there are finite resources available with no new places to put the waste that is generated. His designs aim to make best use of materials and resources to respect the balance between humans and the physical environment. Rau proposes that humans should behave as guests on the planet, rather than owners of it. One way to realise this in building and product design is to shift away from ownership and towards stewardship. Owning products brings responsibility for their upkeep and ultimate disposal, which is problematic when many products are full of materials that are unidentified and difficult to reclaim as an equally valuable element.

As part of the drive towards optimising the use of materials, Rau has implemented the use of service models, and as well as leasing services in some of his buildings he has set up a company called Turntoo that facilitates service-based contracts between manufacturers and users. The idea of Turntoo is that the manufacturer retains ownership of the products and the consumers only pay for the performance, rather than the raw materials that go into the product. This creates a closed loop, with the materials and components of the product remaining in a 'raw materials cycle' while the service offered by the product is agreed with the customer.

Based on this concept, Rau approached Philips lighting and proposed that they offer him a lighting service instead of purchasing lamps and control gear. The idea developed into Philips' 'Pay-per-lux' model, which was pioneered in RAU Architects' own offices in Amsterdam and subsequently in Brummen Town Hall. The Pay-per-lux model is a performance-based arrangement where Philips installs and maintains a lighting level, and monitors the lighting performance and energy use online with annual reporting, health checks and preventative maintenance. Furthermore, Philips pays the energy bills for the lighting, which incentivises it to provide the most efficient lighting system and ensure that it is operating as well as possible.

As Thomas Rau says: 'I told Philips, "Listen, I need so many hours of light in my premises every year. You figure out how to do it. If you think you need a lamp, or electricity, or whatever – that's fine. But I want nothing to do with it. I'm not interested in the product, just the performance. I want to buy light, and nothing else."'[2]

The model has been applied at Schiphol Airport in Amsterdam, and was reconceived by Philips in the UK specifically for the National Union of Students. Now the model has been considered for other products, and pay-per-use lifts look like a promising alternative to the conventional model of capital purchase and a long-term maintenance contract.

Market competition drives down the cost of products, and manufacturers are not incentivised to invest in the durability of the equipment they sell. Even worse, built-in obsolescence is common in consumer goods, driving customers to upgrade products or spend money on maintaining equipment that could have lasted longer.

On the innovative Park 20|20 (see case study in Chapter 13), Coert Zachariasse of Delta Development Group decided to create a new 'product as a service' model to incentivise built-in-durability. Based on his experience of trialling pay-per-use models for various elements such as lighting, carpets and partitions, he concluded that they are best applied to products that have a long life, a high residual value, and high maintenance requirements. This insight led him to choose lifts as the perfect candidate, not least because he would be entering into a long-term maintenance contract with the manufacturer.

He challenged Mitsubishi to create a new business model that entailed Delta paying no capital cost and only paying for the use of the lifts. The aim of the model was for Mitsubishi to retain the ownership and responsibility for the whole life of the lifts, incentivising them to invest in components that are high quality and low maintenance, and are able to be upgraded or remanufactured and used again.

Nine months of negotiations followed, involving KPMG to calculate the financial benefits and Standard Life as the investor that needed to be convinced. The business case was built around the idea that the 'product as a service' model created residual value, so the lift still had value after 40 years of life, whereas the traditional purchase model left zero residual value. With a residual value of 20%, the depreciation of the asset is slower and the accountants can take these lifetime savings from the first year, as shown in Figure 12.01. Alongside this, the regular maintenance payments were no longer required as they were absorbed into the usage charge.

This has resulted in Mitsubishi's unique M-Use®[3] model, where it retains ownership of the lift and aims to extend the life of equipment, and components such as doors, cages and counter-weights are reused at other locations.

By retaining ownership of the product, the theory is that the manufacturer would create a more circular product that is more durable, easier to maintain and upgrade, and with components that can be remanufactured or recycled at end of life. This should reduce the demand for raw materials as well as reducing waste. This theory has been proven in practice, as Mitsubishi has now developed a new product where 95% of the components are not welded or glued. It is aiming for C2C certification and is finding ways to demonstrate the savings. For example, removing the toxic chemicals from paints means they do not have to have specialist areas and the same level of protection for employees.

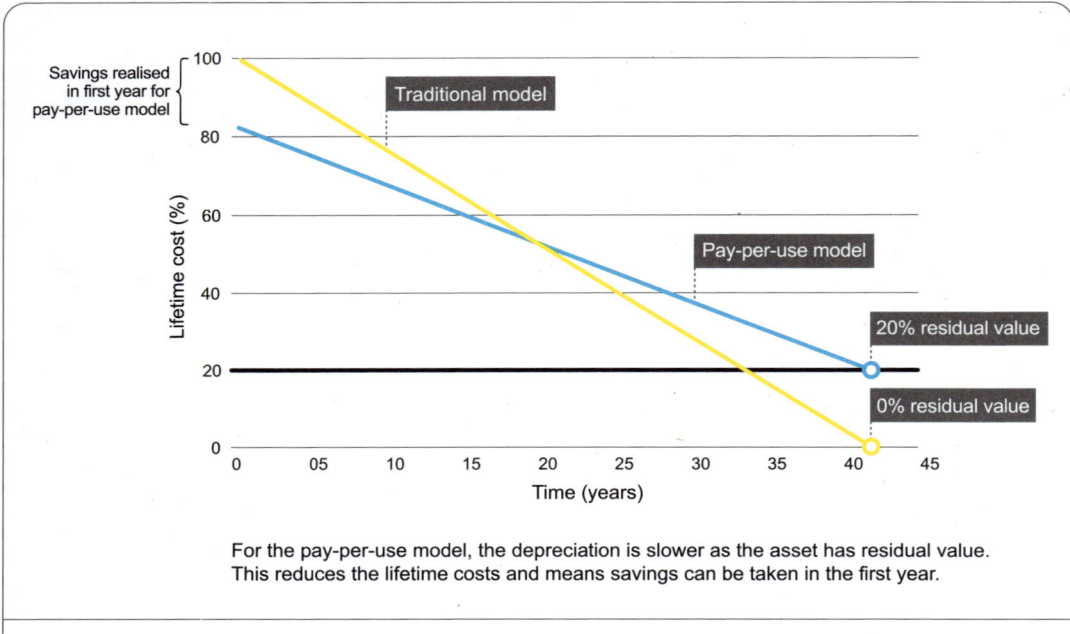

Savings realised in first year for pay-per-use model

Lifetime cost (%)

Traditional model

Pay-per-use model

20% residual value

0% residual value

Time (years)

For the pay-per-use model, the depreciation is slower as the asset has residual value. This reduces the lifetime costs and means savings can be taken in the first year.

Figure 12.01: Pay-per-use model reduces lifetime costs (based on an interview with Mitsubishi and Delta Development Group).

INCENTIVISING RETURN

Selling services instead of products is one way to ensure that building components and contents are returned to the manufacturer for reuse or remanufacture. Another way to close the loop is to incentivise customers to return products when they are no longer required by offering to buy them back or provide a discount on the next service.

Some organisations have recognised the advantages of ensuring that their products are returned, and have set up the infrastructure to collect products, the incentives to ensure a high return rate, and remanufacturing capabilities to squeeze the maximum value out of the used products.

Caterpillar manufactures machinery and engines, including construction and mining equipment. It uses its vendor and distribution system to collect used components, and operates a deposit and discount system to incentivise their return.[4] Customers pay a deposit when they purchase new equipment, which is paid back when they return the product. Caterpillar operates a remanufacturing division that disassembles the products, then cleans and inspects the components to determine what can be salvaged. The components and the 'core' of the machinery or engine can then be remanufactured to a very high performance standard, which allows Caterpillar to offer the same warranty for its 'reman' engines as for new products.[5] This process enables Caterpillar to:

- get valuable feedback on the design of its products
- reduce materials costs
- provide a reliable source of remanufactured components
- provide a parallel business stream of remanufactured products
- build long-term relationships with their customers.

This business model is starting to be applied to building components and contents, such as furniture.

Rype Office, London

Furniture reuse is a well-established practice in the UK, with networks and organisations offering reclaimed furniture. However, WRAP estimates that only 14% of office desks and chairs are reused every year in the UK, with the remainder going to landfill, energy recovery or recycling.[6] This equates to around 75,000 tonnes of office furniture going to landfill annually.[7]

    Rype Office draws on circular economy principles. Its business model addresses every stage in the life of furniture, with the aim of moving beyond

Figure 12.02: Remanufacturing process.

the linear model of consumption and disposal. Rype Office helps companies to refresh and resize their existing furniture, remake furniture sourced from elsewhere at half the cost of new, and choose new furniture that is easy to remanufacture, potentially giving it a much longer life.

It also offers guaranteed buy-back on its furniture and even a leasing service, giving customers the flexibility to return unwanted furniture or to lease more as their organisation changes.

According to Dr Greg Lavery, director of Rype Office, the most common reasons for not using remanufactured furniture are concerns about quality and about the volumes that can be supplied.

'When you put top grade remanufactured furniture alongside new you cannot tell the difference, not even with a magnifying glass – that is how good it is. Modern resurfacing technologies and remanufacturing processes are very high quality. And of course you can choose the colours and finishes that you want because we are completely remaking it,'[8] says Lavery in response to the first concern.

And in response to the second concern about whether the volume of stock is available, he notes that there are hundreds of thousands of furniture items from office clearances now sitting in warehouses since the 2008 global financial crisis, just waiting to be remanufactured. Rype Office's remanufacturing model:

- creates remade furniture at half the cost of new
- reduces the environmental footprint of each piece by an estimated 70%
- creates skilled jobs in the UK
- reduces the amount of new furniture and components that have to be imported into the UK.

CONCLUSION

These service models could be extended to encompass the whole fit-out of a building, allowing occupants to lease everything they need from the manufacturer. This could give occupants access to the latest technology or trends, and enable them to transfer the risks of trialling these new ideas to the manufacturer, along with the total cost of ownership. Zachariasse's insight from testing out various pay-per-use models limits the list of suitable products to those that have a long life, a high residual value and high maintenance requirements.

At first glance, pay-per-use models may appear more expensive than simply purchasing products outright. However, this can change when the total cost of ownership (TCO) is considered. Calculating the TCO involves gathering together a whole range of costs that are not normally associated with the purchasing of equipment. This includes costs associated with maintenance, premature failure, functional obsolescence, replacement and the storage of surplus equipment or replacement parts. Depending on the models, it could also include the energy costs of mechanical and electrical equipment and the disposal costs of products, including hazardous materials that would require specialist handling. The Mitsubishi case study shows that making a more durable product slows depreciation, increases residual value and allows the overall costs to be reduced.

These business models help to ensure that products and materials are kept in circulation for longer and also offer multiple benefits to manufacturers and customers, if they are set up to the benefit of both parties and the right infrastructure is put in place. Manufacturers can develop longer-term relationships with their customers, while reducing their cost of materials and their exposure to volatile materials prices. They can create a second revenue stream and customer base by selling remanufactured products and obtain feedback on their products, allowing them to develop higher-quality, longer-lasting goods. Remanufacturing creates local employment opportunities and reduces the need to import goods. The customers benefit from lower-cost products with the potential to enter into service agreements that mean they are not responsible for the maintenance, upgrade and disposal of products.

VIRTUOUS CIRCLES

If humans were to devise products, tools, furniture, homes, factories, and cities more intelligently from the start, they wouldn't even need to think in terms of waste, or contamination, or scarcity. Good design would allow for abundance, endless reuse, and pleasure

— William McDonough and Michael Braungart

Innovative design and creative thinking need the right environment to flourish. Engaging with the supply chain at the back end of the design process, once all the main decisions have been made, sounds wrong. To then beat the suppliers down to the lowest cost through competitive tendering is not going to create a nurturing environment for new ideas.

Engaging and partnering with the supply chain is standard practice in other industries, leads to strong, trusting relationships, and creates the space to innovate.

Moving away from lowest-cost tendering sounds expensive, but Coert Zachariasse of Delta Development Group, based in Amsterdam, has made it work (see case study below) by implementing an open-book accounting system and by asking the suppliers to provide the best product they can for the allocated budget. In return, they become the guaranteed suppliers for a series of building projects. This creates a fertile environment for design innovation, avoids construction delays and creates a better building that commands more rent.

Zachariasse has also demonstrated at Park 20|20 that it is possible to design buildings using circular economy principles. The buildings are designed to be more adaptable to reduce the risks of obsolescence, the careful selection of materials and components has led to a healthier internal environment, and the buildings should have a residual value by designing for disassembly.

PARK 20|20

Park 20|20 is a new business park one train stop from Amsterdam's Schiphol Airport. It is the first business park in the Netherlands to be inspired by Cradle to Cradle® thinking and aims to implement many of the principles of the circular economy. Park 20|20 is the brainchild of Delta Development Groups' CEO, Coert Zachariasse.

Zachariasse's journey to creating Park 20|20 started in 2004, when he was demolishing buildings on the redundant Fokker airfield factory site in Amsterdam. He realised that there were valuable materials on the site that would be wasted through conventional demolition practices, so he initiated a process whereby he retained ownership of onsite materials which were subsequently deconstructed and reused. Four years later he was inspired by a presentation given by William McDonough on the Cradle to Cradle® philosophy (see Chapter 1) and decided that he wanted to apply the principles to a new business park. He obtained the land, the support and the funding that he needed based on this vision. Figure 13.01 shows the masterplan for the park.

Zachariasse wanted to obtain the right products to enable the entire building to be Cradle to Cradle Certified®, but in meetings with McDonough it quickly became apparent that not enough products were certified at the time to achieve this. Rather than water down his goal, McDonough and Zachariasse developed a pioneering approach to drive the supply chain to respond. The team scanned

Figure 13.01: Park 20 | 20 masterplan.

350 different products for their potential to be Cradle to Cradle® products and rated them red, amber and green. They talked to the amber companies and asked them to phase out particular materials or chemicals in their products to allow them to be certified. This resulted in 41 suppliers who either had C2C certification or who had been vetted to provide an acceptable standard of product.

Zachariasse then changed the whole tendering process to provide a direct relationship with the manufacturers and suppliers. First, he made the contractor a partner in the project and created an open-book accounting system. He asked them to break down their costs into direct costs (materials and labour), site and construction costs, and the profit margin. Zachariasse then offered to double their profit margin in return for total transparency on the direct costs. Between them, they prepared a budget for each of the elements of the building (facades, structure, upper floors, etc.) and then asked the suppliers to provide them with the highest-quality product that they could for the cost. This approach encouraged the suppliers and manufacturers to propose innovative solutions in the early design of the project, while giving Zachariasse control over the supply chain. The manufacturers knew they were the suppliers for the project and so could prepare and manage their supply chains accordingly. This meant that there were no delays or issues with materials and products not being delivered on time, saving time and money on site. BIM was used to create further savings in both construction costs and materials procurement.

Turning the procurement process on its head like this is in sharp contrast to the traditional procurement route, where suppliers are asked to tender for work after all the design decisions have been made. It means that developers, designers and contractors can work collaboratively with the manufacturers to create new solutions that use fewer materials, waste less and provide quality solutions.

Overall, the shortened construction time and reduced supply chain costs saved 18% on the project. The same team has been retained for the other buildings on the business park, which has given them security, while keeping the consistency and allowing them to learn and improve the design. Many of the same building elements, supply chain and detailing is used on the new buildings, with refinements in the design as lessons are learned. This helps to address one of the fundamental problems with building design and construction: each building is an untested prototype. By integrating the supply chain and repeating the design, the lessons from the first building are fed into the next and the design effort is not wasted. This makes it sound like all the buildings will be the same, but of course this is not the case as the building system allows for different facades to be used, different configurations of spaces and different geometries, as shown in Figure 13.02.

Delta Development Group is moving beyond the traditional developer role. It works closely with the tenants to develop spaces that match what they actually

Figure 13.02: ANWB Reizen building with part of Bosch Siemens building in the foreground.

Figure 13.03: Bluewater building, interior.

need rather than what it thinks they want. Zachariasse says that 'real estate starts with metrics such as floor area, when it should be starting with values and a vision, developing goals, objectives and then settling on metrics'.[1] When Bluewater said it wanted 11,000 m² of office space on the 20|20 Park, Delta discussed this and they worked out that their current working models meant they needed around 6,000 m². They settled on 8,000 m². Zachariasse notes that the role of offices is changing; they are now for meeting colleagues and places to find inspiration rather than just places for desk-based work. This means that organisations often need less space, or different types of space (see Figure 13.03).

Delta also fits out the building to provide a 'turnkey' solution, as this gives it control over the choice of materials for the interiors. As part of the fit-out, Delta has entered into leasing arrangements for some of the elements. There are leasing contracts with LED Lease and BB Lights for the lighting, office furniture is leased from Ahrend, the carpet tiles are leased from Desso and there is a pay-per-use contract with Mitsubishi lifts (see Chapter 12). Delta is also working with HM Ergonomics, which is researching a full interior leasing concept, including partition walls.

The results are impressive. The buildings at Park 20|20 are all fully let on completion, and Zachariasse's figures show that they are commanding 'a 70–80% higher rental than the buildings on the other side of the dyke'.[2] The buildings:

- are designed with flair, despite (or perhaps because of) the limited palette
- have abundant daylight
- have plenty of internal plantings
- are constructed with non-toxic materials.

Large, central atriums provide plenty of space for circulation, breakout areas from meetings and events, and a sense of space and light. Internal green walls and trees provide contact with nature, enhance air quality and bind fine dust (see Figure 13.04).

The healthy indoor environment is helping Delta Development Group attract tenants, as well as commanding higher rents. Zachariasse has created some interesting performance-based leasing models. He starts by offering an incentive package to new tenants formed of a rent reduction of 30%, but the amount of incentive is based on the results of the occupant satisfaction survey.

The idea is that the tenants are paying a premium for a healthy, productive workspace, and if this positive impact is delivered then they will be happy to pay the full rent. Delta has set up an arrangement to objectively measure the occupant satisfaction using the Leesman Index, which is a benchmarking tool that measures workplace effectiveness. It captures employees' feedback on how well the workplace environment supports their activities. Delta uses the benchmarks to create a sliding scale: if the occupant survey achieves a high satisfaction rating, then Zachariasse does not have to give any incentives; if they achieve a lower score, then they have to reduce the rent. The idea, according to Zachariasse, is that 'we let clients/employees judge our work and add an economic model to recognise this'.

The building has certainly delivered on wellbeing and productivity. When Bosch Siemens employees were surveyed by Delta both before and after they moved into the new office, the results of the second survey showed a 5–6% increase in productivity.

The buildings have been consciously designed to allow adaptation to extend the life of the building and for disassembly at end of life. Notably, the designers have attempted to solve one of the intractable problems when designing for adaptation and disassembly: how to construct the upper floors. Park 20|20 has pioneered a flooring system called Slimline™, which consists of a precast concrete ceiling integrated with steel beams and a sub-floor that allows access to the service void created by the steel beams (see Figure 13.05).

The accessible horizontal void allows services to be upgraded to the latest technology without having them poured in concrete. This design feature should also allow for the building to be sufficiently adaptable to accommodate different tenants. When the market changes and offices need to be turned into a different type of building, say a hotel or apartments, the system is designed to allow this adaptation by providing:

Figure 13.04: Bosch Siemens building atrium, with green wall.

- the potential to add (or remove) services within the floor void
- acoustic separation using double-layer construction
- modular design that allows staircases to be reconfigured by removing whole floor sections without having to break up screed
- a precast finished soffit that has a smooth finish for aesthetic value.

The system, as the name suggests, provides a slimmer floor sandwich than conventional construction and uses considerably fewer materials, providing a weight saving for the whole building. At end of life, the steel beams are relatively easy to separate for recycling and the concrete can be, inevitably, downcycled for aggregate.

The C2C certification includes criteria to assess whether the products can be demounted or disassembled (see Chapter 9). This has been used in Park 20|20 to provide buildings that can be redesigned with a new appearance as well as a new function. The Bosch Siemens building includes a demountable, Cradle to Cradle Certified® ceramic tile facade from Mosa that can be removed and replaced, if a new look is required.

The buildings are designed around an 8 x 8 m grid, which works equally well for commercial or residential buildings in the Netherlands. The structural steel frame is modular and uses standard-sized beams to increase the chances of them being reused at end of life.

Since its construction, the Bosch Siemens building has had a minor renovation

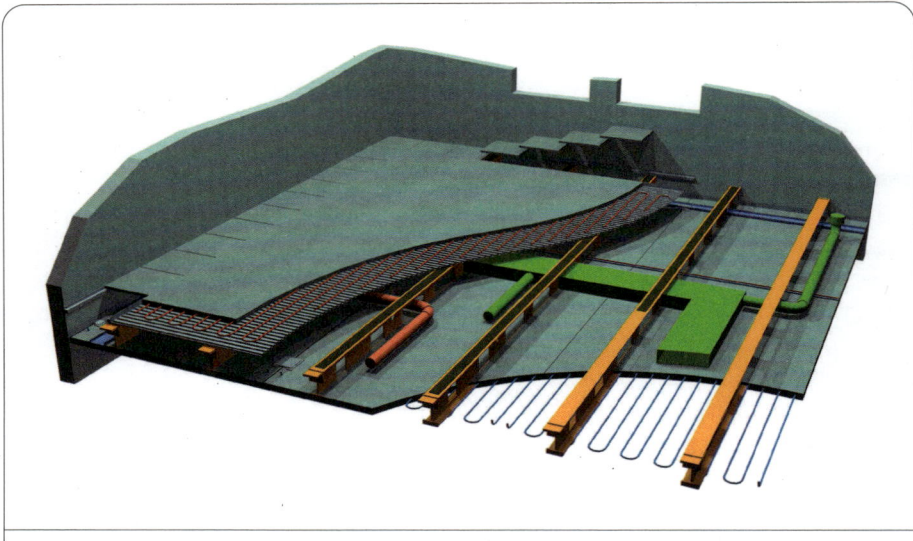

Figure 13.05: Slimline™ flooring system.

and retrofit of the showroom, reception area and part of the office. The circular economy principles were tested during the retrofit and the interior walls and acoustic panels were successfully reclaimed and used in another building.

Delta asked a demolition contractor to provide a quote for reclaiming the materials on the conventional building design of the Bluewater building on the 20|20 Park. Zachariasse's figures show that there is €1 million-worth of steel in the building. Using conventional construction techniques the contractor quoted €45,000-worth of scrap metal and €48,000 to extract the metal from the building, leaving a negative residual value of -€3,000 to reclaim the steel for scrap. This is an example of how buildings typically have a negative residual value at end of life. The use of a structural screed was cited as the main reason for the high cost of extraction and the low value of the salvaged steel. Once the designers had redesigned the building to make it easier to deconstruct, the value of the steel went up by €195,000, with an additional construction cost of €70,000 and additional deconstruction costs of €45,000. This left a net residual value of €85,000, giving the building a potential positive residual value at end of life.

CONCLUSION

Park 20|20 demonstrates that buildings really can be designed for a more circular economy, and that the principles discussed in this book can be implemented. The case study shows that designing within a circular economy makes financial sense but, to really work, the construction industry needs to be turned upside down and given a shake.

Designing for adaptability or deconstruction is hard to justify and is unlikely to happen unless it is part of a wider story that starts with reducing construction time on site, continues with the ability to retain value by adapting buildings to changing markets and concludes with the attractive idea of providing residual value rather than demolition costs.

Figure 13.06 summarises the potential changes in revenue and costs associated with designing buildings with a long-term view, based on the ideas proposed by Zachariasse and inspired by a graph drawn by researchers at Loughborough University in relation to adaptable buildings.[3] It compares the conventional building model with a more adaptable one that is designed for disassembly and uses materials that provide a healthier internal environment. The collaborative approach used by Delta Development Group can reduce construction time, and therefore cost, while the costs of refitting and adapting the spaces, or even the building, should be lower and require less time. There is the potential to have lower rates of depreciation and perhaps to sustain higher rental yields. Finally, designing for disassembly can provide a positive residual value.

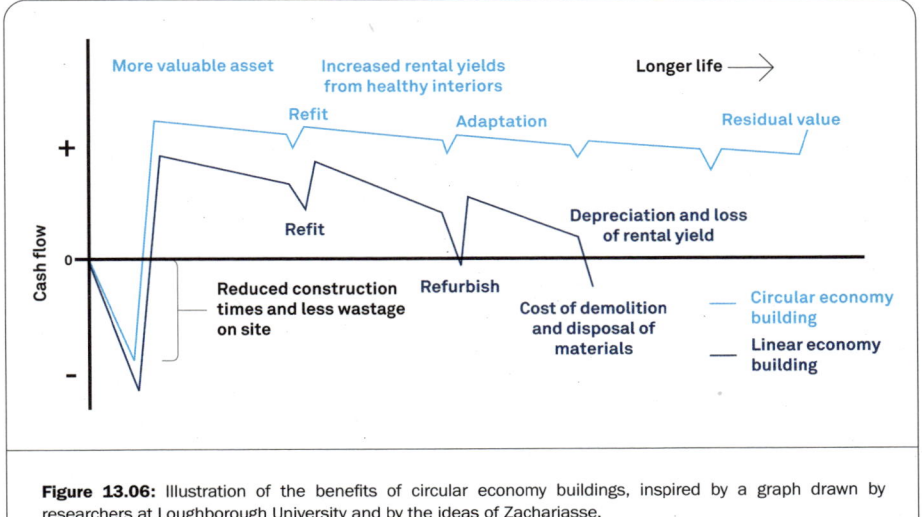

Figure 13.06: Illustration of the benefits of circular economy buildings, inspired by a graph drawn by researchers at Loughborough University and by the ideas of Zachariasse.

Making small, incremental changes towards a more circular economy could drive up capital costs or design fees, which can be hard to defend against the onslaught of lowest-cost tendering and value engineering. Embracing the whole circular economy narrative, on the other hand, means collaboration with the entire supply chain and the ability to design for the whole life of the building, including 'beyond the grave'.

Buildings are a legacy for future generations. There is an opportunity to leave a valuable inheritance of buildings that can be adapted to new uses or deconstructed and turned into new buildings, in whatever form the future demands.

COMING
FULL CIRCLE

You don't have to do any of this; survival is not mandatory.

— Walter R. Stahel

This book demonstrates that circular economy thinking can be applied to buildings, and that there are opportunities for those who embrace the ideas to create new business models and designs that enable a transition to a more regenerative built environment.

The cases studies in this book show that there are some visionary and enterprising clients, designers, constructors, manufacturers and entrepreneurs who are capitalising on the opportunities presented by a circular economy. They have managed to overcome the barriers to change and the inertia in the system to differentiate themselves in the market.

These new ideas address the pressing need to respond to the challenge of the climate crisis by slowing and reversing the growing demand for raw materials, and eliminating the mountains of waste produced every year. By implementing these ideas, businesses can seize more control over their raw materials and protect themselves from supply disruptions and volatile prices, while reducing their exposure to environmental legislation and the costs of waste disposal. Businesses can respond to the shift in the market by providing performance rather than products, allowing them to forge new longer-term relationships with their customers.

CYCLING, NOT RECYCLING

Embracing the idea of a circular economy is so much more than just reducing waste arising on site, or recycling more waste. It is about fundamentally rethinking the way that buildings are designed, procured and used. It is about avoiding design that takes buildings down blind alleys, where there is no way back except depreciation, demolition and downcycling. It is about creating a built environment that retains the value of buildings, components and materials in a virtuous circle, while making buildings that are better for the environment and for people.

THINKING BEYOND THE GRAVE

This book outlines a set of design principles that interpret the circular economy model for the built environment. The trick is to think beyond the immediate demands for the building and consider the potential next lives of the building and its components. The design principles aim to capture circular economy thinking and to apply it to building design, as follows:

- Design-out waste by refitting and refurbishing buildings in preference to demolition and building new, using waste as a resource to create new designs and using lean design principles to create buildings that use fewer resources and components and are less complex.

- Create structures that are built to last, have layers that can be peeled off and replaced, can be adapted to new uses and can be disassembled and even reconfigured for different uses on new sites.
- Select building components that are carefully designed to flow in either a technical or a biological cycle, depending on their lifespan, use and what is available. The components in the technical cycle are designed to retain value, either by being designed for reuse, remanufacture or disassembly. The biological materials remain uncontaminated and can be cascaded through different uses before being returned to the biosphere at end of life. Building components and materials in the technical cycle that are hard to salvage at end of life should be designed-out or perhaps replaced with biological materials.

LEAPS AND BOUNDS

New technologies and business models can leapfrog over current thinking and leave it standing, allowing industries to escape the bounds of convention.

3D printing, or additive manufacture, provides the potential to make beautiful components that do not need complex polymers or irreversibly bonded materials. New designs could be inspired by the way that nature uses structure and form to create stiffness, flexibility and even colour from its limited palette of materials. Additive manufacture reduces waste during production and, if applied correctly, could help to create components made from only a few polymers that are simpler to recycle at end of life.

The drive towards DfMA is critical to implementing lean design principles, cutting resource use and eliminating site waste. Turning from construction to assembly on site will help to eliminate wet trades, and provides the opportunity to also design for disassembly. DfMA should become DfMAD, where the second 'D' is for disassembly.

Information technology is being used to locate potential sources of materials via online exchanges and resource maps. The materials and components in buildings can be fully catalogued using BIM so they can become materials banks for future generations. These inventories can be used to create a market for components within a building before it is demolished, which could be an incentive for it to be deconstructed rather than demolished, allowing items to be salvaged.

New business models show how thinking differently about ownership can enable valuable resources to be kept in high-value cycles, where manufacturers retain ownership of the product or incentivise return to ensure that components and materials are kept in circulation. There is also the potential for new local industries to be set up to remanufacture or reprocess products that would otherwise have been treated as waste.

The Park 20|20 case study shows that it is possible to bring many of these design principles and concepts together to create buildings that are not only constructed from certified products and designed for adaptation and deconstruction, but also cost less to build, command higher rents and provide healthier internal environments.

The collaborative approach used by Delta Developments allows it to have a direct relationship with the manufacturers and suppliers, which in turn enables them to focus on materials and products selection, and to nurture innovative design solutions.

There is no need to wait for the policy landscape to shift or for the systemic problems in the industry to be resolved. With the right mindset, individuals and organisations can embrace this new agenda and create better buildings that leave a positive legacy for future generations.

ABOUT THE AUTHOR

1. The London Plan 2021 https://www.london.gov.uk/what-we-do/planning/london-plan/new-london-plan/london-plan-2021 (accessed June 2021).

INTRODUCTION

1. Ellen MacArthur Foundation, 'Towards the Circular Economy: Economic and Business Rationale for an Accelerated Transition', 2013.
2. Global Status Report, 'Towards a Zero-emission, Efficient, and Resilient Buildings and Construction Sector', UN Environment, 2017.
3. OECD, 'Materials Resources, Productivity and the Environment', *OECD Green Growth Studies*, Paris: OECD Publishing, 2015, pp 64, 84.
4. 'For many metals, this means that about three times as much material needs to be moved for the same quantity of metal extraction as a century ago'; 'The tendency to process lower grades of ore to meet increasing demand is leading to a higher energy requirement per kilogram of metals, and consequentially to increased production costs', UNEP International Resource Panel, 'Decoupling 2: Technologies, opportunities and policy options', UNEP, 2014.
5. 'The extraction and use of material resources is closely linked to negative impacts on aspects of the environment', UNEP, 'Decoupling 2'.
6. Ellen MacArthur Foundation, *Towards the Circular Economy*.
7. 'Cement technology roadmap plots path to cutting CO_2 emissions 24% by 2050', 6 April 2018, https://www.iea.org/news/cement-technology-roadmap-plots-path-to-cutting-co2-emissions-24-by-2050 (accessed 11 December 2020).
8. 'Mission Possible: Reaching Net-zero Carbon Emissions from Harder-to-abate Sectors by Mid-century,' 2018.

CHAPTER 1

1. R.U. Peiró, 'Material efficiency: rare and critical metals', *Philosophical Transactions*, The Royal Society, 2013.
2. Ellen MacArthur Foundation, 'Towards the Circular Economy: Economic and Business Rationale for an Accelerated Transition', 2013.
3. House of Commons Environmental Audit Committee, 'Growing a circular economy: Ending the throwaway society, Third report of session 2014–15', London: The Stationery Office, 2014.
4. European Commission, 'Moving towards a circular economy', 2014, http://ec.europa.eu/environment/circular-economy/index_en.htm (accessed 8 Febraury 2021).
5. Ellen MacArthur Foundation, 'Delivering the Circular Economy: A Toolkit for Policymakers', 2015.
6. Walter R. Stahel, 'Product Life Factor', Product Life Institute, 1982, http://www.product-life.org/en/major-publications/the-product-life-factor (accessed 8 February 2021).
7. Michael Braungart and William McDonough, *Cradle to Cradle: Remaking the Way We Make Things*, New York: North Point Press, 2002.
8. *Ibid.*, p 72.
9. *Ibid.*, p 76.
10. *Ibid.*, p 75.
11. *Ibid.*, p 61.
12. *Ibid.*, p 105.
13. Michael Pawlyn, *Biomimicry in Architecture*, second edition, London: RIBA Publishing, 2016.
14. Ellen MacArthur Foundation, *Towards the Circular Economy*.
15. CIBSE, 'Resource Efficiency of Building Services', London: CIBSE, 2013.
16. https://www.circularonline.co.uk/news/lamp-luminaire-recycling-rates-still-rise/ (accessed June 2021)
17. David H. Clark, *What Colour is Your Building?* London: RIBA Publishing, 2013, p 51 (a pie chart shows an example breakdown of construction embodied carbon in a new office building, with 13% and 42% associated with the substructure and superstructure, respectively).
18. Bioregional, 'Reclamation led approach to demolition', London: Defra, 2007.
19. World Economic Forum, 'Towards the circular economy: accelerating the scale-up across global supply chains', Geneva: World Economic Forum, 2014.
20. *Ibid.*

CHAPTER 2

1. 'Global Resources Outlook: Natural Resources for the Future We Want', UN Environment International Resource Panel, 2019.
2. Material resources are biomass, fossil fuels, metals and non-metallic minerals used in the economy.
3. 'Global Material Resources Outlook to 2060: Economic Drivers and Environmental Consequences', OECD Publishing, Paris, 2019, https://doi.org/10.1787/9789264307452-en.
4. Global Status Report 2017, 'Towards a Zero-emission, Efficient, and Resilient Buildings and Construction Sector', UN Environment, 2017.
5. IRP and UNEP, *The Weight of Cities: Resource Requirements of Future Urbanization*, 2018.
6. Ellen MacArthur Foundation, 'Completing the Picture: How the Circular Economy Tackles Climate Change', 2019.
7. Ellen MacArthur Foundation, *Towards the Circular Economy*.
8. CIBSE, *Resource Efficiency of Building Services*.
9. Marc Schmid, 'Rare Earths in the Trade Dispute Between the US and China: A Déjà Vu', *Intereconomics*,

54(6), 2019, https://www.intereconomics.eu/contents/year/2019/number/6/article/rare-earths-in-the-trade-dispute-between-the-us-and-china-a-deja-vu.html (accessed 17 January 21).

10. UNEP International Resource Panel, 'Decoupling 2'.

11. Circular Economy Task Force, 'Resource resilient UK', London: Green Alliance, 2013.

12. Julian Allwood and Jonathan Cullen, *Sustainable Materials: With Both Eyes Open*, UIT Cambridge, 2012, p 6.

13. House of Commons Environmental Audit Committee, 'Growing a circular economy'.

14. Circular Economy Task Force, 'Resource resilient UK'.

15. European Commission, 'Moving towards a circular economy', 2014, http://ec.europa.eu/environment/circular-economy/index_en.htm (accessed 8 February 2021).

16. Prism Environment, 'Construction Sector Overview in the UK', EISC Ltd, 2012, p 9.

17. J.M. Allwood, et al., 'Absolute zero: Delivering the UK's climate change commitment with incremental changes to today's technologies', UK FIRES, University of Cambridge, 2019.

18. CIBSE, 'Resource efficiency of building services'.

19. Department of Environment, Food and Rural Affairs, 'UK Statistics on waste', Government Statistical Service, 2020.

20. *Ibid.*

21. All-Party Parliamentary Sustainable Resource Group, 'Remanufacturing towards a resource efficient economy', London: All-Party Parliamentary Sustainable Resource Group, 2014.

22. House of Commons Environmental Audit Committee, 'Growing a circular economy'.

23. Ellen MacArthur Foundation, *Towards the Circular Economy*.

24. Chatham House, 'A global redesign? Shaping the circular economy', London: Chatham House, 2014.

25. Ellen MacArthurFoundation, *Towards the Circular Economy*.

26. The London Plan 2021 https://www.london.gov.uk/what-we-do/planning/london-plan/new-london-plan/london-plan-2021 (accessed June 2021).

CHAPTER 3

1. J. O'Connor, 'Survey of actual service lives for North American Buildings', Canada: Forintek Canada Corp, 2004.

2. Habraken, *The Structure of the Ordinary*, p 6.

3. Forest Lee Flager, 'The design of building structures for improved life-cycle performance', Cambridge, MA: The MIT Press, 2003.

4. Junko Edahiro, 'Rebuilding Every 20 Years Renders Sanctuaries Eternal – the Sengu Ceremony at Jingu Shrine in Ise', http://www.japanfs.org/en/news/archives/news_id034293.html, August 2013 (accessed 8 February 2021).

5. Colleen McKeracher, 'Designing for Destruction: Anticipating Architectural Dismantling Through the Act of Making', Carleton University Ottawa, 2014.

6. A.M. Baum and A. McElhinney, 'The causes and effects of depreciation in office buildings: A ten year update', 1996.

7. *Ibid.*, p 16.

8. S.J. Wilkinson, H. Remoy and C. Langston, *Sustainable Building Adaptation: Innovations in Decision-Making*, London: RICS, 2014, section 2.4.3.

CHAPTER 4

1. Telephone interview between Howard Button and David Cheshire, 3 October 2014.

2. W. Addis and J. Schouten, *Principles of Design for Deconstruction to Facilitate Reuse and Recycling*, London: CIRIA, 2014.

3. *Waste Duty of Care Code of Practice*, London: DEFRA, November 2018, section 2.4.

4. https://urbanflows.ac.uk/generate/ (accessed June 2021).

CHAPTER 5

1. The London Plan 2021 https://www.london.gov.uk/what-we-do/planning/london-plan/new-london-plan/london-plan-2021 (accessed June 2021).

2. https://www.leti.london/publications (accessed 4 December 2020).

3. https://www.architecture.com/about/policy/climate-action/2030-climate-challenge (accessed 4 December 2020).

CHAPTER 6

1. Stewart Brand, *How Buildings Learn – What Happens After They're Built*, London: Phoenix, 1994, p 13.

2. Julian Allwood and Jonathan Cullen, *Sustainable Materials*, p 242.

3. R. Hartwell and M. Overend, 'Unlocking the Re-use Potential of Glass Façade Systems', Proceedings of Glass Performance Days, June 2019, Finland.

CHAPTER 7

1. Aldersgate Group, *Resource Efficient Business Models: The Roadmap to Resilience and Prosperity*, London: Aldersgate Ltd, 2015.

2. Julian Allwood and Jonathan Cullen, *Sustainable Materials*, p 247.

3. 'Reducing and recycling plasterboard waste on a site where space is constrained', London: WRAP, n.d.

4. 'Tate Modern 2, case study,

designing out waste', London: WRAP, 2010.
5. The Institute of Civil Engineers, 'ICE Demolition Protocol', ICE, 2008.
6. 'Circular Economy Statement Guidance', Draft for consultation, Greater London Authority, October 2020.
7. 'Circular economy how-to guide: reusing products and materials in built assets', UK Green Building Council, April 2020.
8. Bioregional, 'Reuse and recycling on the London 2012 Olympic Park: Lessons for demolition, construction and regeneration', London: Bioregional, 2011.
9. 'Fit for (re)purpose: The environmental and financial benefits of a circular and fitted approach to leasing office space. A comparative case study', Paul Jaffe, dissertation, University of Cambridge, 2020.
10. Ibid.
11. T. Cooper, Longer Lasting Products: Alternatives to the Throwaway Society, Farnham: Gower Publishing Limited, 2010.
12. https://www.surveyor-services.co.uk/what-are-the-grade-a-office-specifications/ (accessed 15 February 2021).
13. 'Projects worth £600 billion in the pipeline as government gets Britain building', GOV.UK, https://www.gov.uk/government/news/projects-worth-600-billion-in-the-pipeline-as-government-gets-britain-building (accessed 9 February 2021).
14. 'Aiming for the UK's first net zero carbon commercial development', Landsec, https://landsec.com/insights/future-trends/development/aiming-uks-first-net-zero-carbon-commercial-development (accessed 9 February 2021).

CHAPTER 8

1. W. Addis and J. Schouten, Principles of Design for Deconstruction.

2. John Habraken, Supports, an Alternative to Mass Housing, London: The Architectural Press, 1972, translated by B. Valkenburg.
3. Working Commission W104 Open Building Implementation, http://www.open-building.org/ob/concepts.html (accessed 9 February 2021).
4. Denise Morado Nascimento, 'N.J. Habraken explains the potential of the Open Building approach in architectural practice', Vitruvius, http://www.vitruvius.com.br/revistas/read/entrevista/13.052/4542?page=3 (accessed 9 Febraury 2021).
5. F. v. Gassel, 'Experiences with the Design and Production of an Industrial, Flexible and Demountable (IFD) Building System', 2002.
6. Julian Allwood and Jonathan Cullen, Sustainable Materials, p 228.
7. Stewart Brand, How Buildings Learn, p 12.
8. D. Gazard, 'Abbey Mill', Bradford on Avon Museum Booklet, 2012.
9. TEC Architecture and 3DReid, 'Adaptable design: Adapt and survive: A proposal for a form based design code', London: TEC Architecture and 3D Reid, n.d.
10. Telephone interview with Chris Gregory, TEC Architecture Limited, 2 March 2015.
11. Grimshaw brochure, School of Art and Design, Bath Spa University UK, 10 June 2020.
12. 'A Statement of Expectations, 1976', from Ideas Journal, 1979, pp 18–19.
13. School of Art and Design, Bath Spa University UK, Grimshaw brochure, 10 June 2020.

CHAPTER 9

1. W. Addis and J. Schouten, Principles of Design for Deconstruction.
2. Arup and CIOB, 'Designing for the deconstruction process', Final report, Arup and CIOB, 2013.

3. W. Addis and J. Schouten, Principles of Design for Deconstruction.
4. Paola Sassi, 'Closed Loop Material Cycle Construction', Cardiff: Cardiff University, School of Architecture, 2009.
5. BRE, Dealing with Difficult Demolition Wastes: A Guide, Watford: IHS BRE Press, 2013.
6. WellMet 2050, University of Cambridge, 'Novel Jointing Techniques to promote deconstruction of buildings', Cambridge: WellMet 2050, 2010.
7. Robin Powell, 'An investigation into the potential opportunities and challenges for the demolition sector associated with the demolition of "Offsite" "Manufactured Buildings"', University of Wolverhampton, 2019.
8. Frank Heinlein, Recyclable by Werner Sobek, Werner Sobek, 2019.
9. Email correspondence with Jouke Post, 21 May 2015.
10. 'Project Jack – Google's answer to the problem of flexible spaces', Dr Kerstin Sailer, UCL, August 2016.
11. 'Jack FM Manual', AHMM.
12. 'Project Jack', Dr Kerstin Sailer.

CHAPTER 10

1. Paola Sassi, 'Closed Loop Material Cycle Construction'.
2. Ibid.
3. Ibid.
4. CIBSE, 'Resource Efficiency of Building Services'.
5. Colin M. Rose, et al., 'Cross-Laminated Secondary Timber: Experimental Testing and Modelling the Effect of Defects and Reduced Feedstock Properties', Sustainability, 10(11), November 2018.
6. Ibid.
7. Colin M. Rose and Julia A. Stegemann, 'Material gains', in UN at 75: Sustainable Engineering in Action, Artifice Press, 2020.
8. Jean Michel Leban, et al., 'Wood welding: A challenging alternative to conventional wood gluing',

Scandinavian Journal of Forest Research, 20, 2005, pp 534–538.
9. Michael Braungart and William McDonough, *Cradle to Cradle*, p 98.
10. Julian Allwood and Jonathan Cullen, *Sustainable Materials*, p 53.
11. 'Structural steel reuse, assessment, testing and design principles', Steel Construction Institute, 2019/
12. 'Building Insulation Foam Resource Efficiency Action Plan', London: WRAP, 2012.
13. RWM, *The Circular Economy: Exploring the Potential for your Business*, i2i, 2014.
14. Michael Pawlyn, *Biomimicry in Architecture*, p 35.
15. *Ibid*.
16. Julian Allwood and Jonathan Cullen, *Sustainable Materials*, p 194.
17. 'Guidance on re-use and recycling of used carpets and environmental considerations for specifying new carpet', London: WRAP, 2014.
18. *Ibid*.
19. A. Pearson, 'Fit for purpose', *RICS Modus*, February 2012, pp 40–41.
20. Michael Pawlyn, *Biomimicry in Architecture*, p 39.
21. *Ibid*.
22. 'natureplus e.V. Award Guideline GL0000 Basic Criteria', natureplus, May 2011.
23. 'Cradle to Cradle® Certified™ Product Standard', Version 3.0, MBDC, 2013.
24. 'Banned Lists of Chemicals, Cradle to Cradle® Certified™ Product Standard, Version 3.0', McDonough Braungart Design Chemistry, LLC, 2012.
25. William McDonough + Partners, 'Sustainability Base: NASA's first space station on Earth', San Francisco: William McDonough + Partners, 2012.

CHAPTER 11

1. Interview with Thomas Rau, 17 November 2014.
2. F. Krausmann, et al., 'Global socioeconomic material stocks rise 23-fold over the 20th century and require half of annual resource use', *Proceedings of the National Academy of Sciences*, 114(8), 2017, pp 1880–1885, https://doi.org/10.1073/PNAS.1613773114.
3. A. Ajayebi, et al., 'Spatiotemporal model to quantify stocks of building structural products for a prospective circular economy', *Resources, Conservation and Recycling*, 162, 2020, 105026. DOI: 10.1016/J.RESCONREC.2020.105026.
4. K. Zhou, et al., 'Developing advanced techniques to reclaim existing end of service life (EoSL) bricks – An assessment of reuse technical viability', *Developments in the Built Environment*, 2, 2020, 100006. DOI: 10.1016/j.dibe.2020.100006.
5. T. Kay and J. Essex, 'Pushing Reuse: Towards a Low-carbon Construction Industry', London: BioRegional, 2008, https://bioregional.com.au/wp-content/uploads/2015/05/PushingReuse.pdf (accessed 10 February 2021).
6. Cement Industry Federation, www.cement.org.au, (accessed 10 February 2021).
7. Chatham House, 'Innovation in low-carbon cement and concrete', https://www.chathamhouse.org/about/structure/eer-department/innovation-low-carbon-cement-and-concrete# (accessed 10 February 2021).
8. Sabbie A. Miller, et al., 'Impacts of booming concrete production on water resources worldwide', *Nature Sustainability*, 1, 2018, pp 69–76.
9. 'Why the world is running out of sand', BBC Future, 18 November 2019, https://www.bbc.com/future/article/20191108-why-the-world-is-running-out-of-sand (accessed 10 February 2021).
10. J.M. Allwood, et al., 'Absolute zero, Delivering the UK's climate change commitment with incremental changes to today's technologies', *UK FIRES*, University of Cambridge, 2019.
11. 'UK Concrete and Cement Industry Roadmap to Beyond Net Zero', Minerals Products Association, 2020.
12. 'Super Circular Estate project, Journal No. 4, Project led by the Municipality of Kerkrade', Elma Durmisevic, March 2020.
13. 'UKGBC Future Leaders: Innovation Project Summaries', UK Green Building Council, 2014.
14. Telephone interview with Howard Button, NFDC, 14 September 2020.

CHAPTER 12

1. Walter R. Stahel, from a summary of 'The Performance Economy', www.palgrave.com, 2010 (accessed 15 October 2015).
2. Philips, 'Case study: RAU Architects', Koninklijke Philips Electronics N.V., 2012.
3. https://www.mitsubishi-elevators.com/m-use/ (accessed 10 February 2021).
4. Ellen MacArthur Foundation, *Towards the Circular Economy*.
5. *Ibid*.
6. 'Benefits of reuse case study: Office furniture', London: WRAP, 2011.
7. WRAP estimate, private correspondence between WRAP and Rype Office, 2015.
8. Interview with Greg Lavery, Rype Office, 2 March 2015.

CHAPTER 13

1. Interview with Coert Zachariasse, 17 November 2014.
2. *Ibid*.
3. A. Manewa, et al., 'A paradigm shift towards whole life analysis in adaptable buildings', Changing Roles: New Roles; New Challenges, Noordwijk aan Zee, pp 5–9, October, 2009.

Figure 1.01: Walter R. Stahel, Product-Life Factor, a Mitchell Prize Winning Paper (1982). Redrawn by AECOM.

Figure 1.02: Ellen MacArthur Foundation circular economy team drawing from Braungart & McDonough and Cradle to Cradle (C2C).

Figure 1.03: Ellen MacArthur Foundation.

Figure 3.01 & 3.02: Richard Hind.

Figure 6.01: AECOM - adapted from Stewart Brand, Shearing Layers of Change (1994).

Figure 6.02: David Cheshire, AECOM - adapted from the DEGW 7 'S' model.

Figure 6.05: Rebecca Hartwell.

Figure 6.06: Arup.

Figure 7.01 & 7.02: Matt Chisnall / courtesy of Derwent London.

Figure 7.03: CGI by Cityscape / courtesy of Derwent London.

Figure 7.04 & 7.05: Make Architects.

Figure 7.06 & 7.07: Givesha / courtesy of Elina Grigoriou.

Figure 7.08: Landsec / Bryden Wood Technology.

Figure 7.09: Chris Hill, Unity Works Board.

Figure 8.02: Drawing: AECOM; based on the Multispace research by 3DReid.

Figure 8.03 & 8.05: 3DReid.

Figure 8.06: Grimshaw architects.

Figure 8.07: Paul Raftery

Figure 8.08: Chris Wakefield.

Figure 9.01: By Permission of Michael Samson, SCI.

Figure 9.02: Jie Yang.

Figure 9.03: Werner Sobek.

Figure 9.04: Jouke Post.

Figure 9.05: Luuk Kramer.

Figure 9.06: AHMM.

Figure 9.07: AHMM, credit Tim Soar.

Figure 10.1: GatorDuct.

Figure 10.02 & 10.03: Architype.

Figure 10.04 & 10.05: reproduced by permission of Ehab Sayed on behalf of Biohm.

Figure 10.06: Christopher Pendrich (photographer), reproduced by permission of Sean and Stephen Ltd (architects), www.seanandstephen.com.

Figure 10.07: Kenoteq.

Figure 10.08, 10.09 & 10.10: Photograph: Cesar Rubio / courtesy William McDonough + Partners

Figure 11.01: Dr Danielle Tingley, redrawn by AECOM.

Figure 11.02: Atta Ajayebi, Bradford University.

Figure 11.03: By permission of Kan Zhou and Han-Mei Chen.

Figure 11.04: By permission of Sven Sven Hiskemuller van der Zijden, Rutte Group.

Figure 11.05: By permission of Elma Durmisevic.

Figure 11.06: UK Green Building Council. Redrawn by AECOM.

Figure 11.07: RMF.

Figure 12.01: Based on a sketch by Delta Developments, drawn by AECOM.

Figure 12.02: Rype Office.

Figure 13.01, 13.02 & 13.04: Reproduced by permission of DDG (Delta Developments Group) 2015.

Figure 13.03: Photograph: Sander van der Torren Fotografie, www.torren.nl. Reproduced by permission of DDG (Delta Developments Group) 2015.

Figure 13.05: Slimline Buildings.

Figure 13.06: Adapted from Manewa et al, A Paradigm Shift Towards Whole Life Analysis in Adaptable Buildings (2009). Redrawn by AECOM.

All other images are provided by the author.

Handbook to Building a Circular Economy is printed on Revive 100 Silk Recycled 350gsm and Revive 100gsm paper, FSC Recycled 100% post-consumer waste. Papers Carbon Balanced by the World Land Trust. Printed with eco-friendly high-quality vegetable-based inks by Pureprint Group, the world's first carbon neutral printer.

Published by RIBA Publishing,
66 Portland Place, London, W1B 1AD

ISBN 978 1 85946 954 5

British Library Cataloguing-in-Publication Data
A catalogue record for this book is available from the British Library.

Commissioning Editor: Elizabeth Webster
Assistant Editor: Clare Holloway
Production: Richard Blackburn
Designed by CHK Design
Typeset by Fakenham Prepress Solutions, Norfolk
Printed and bound by Pureprint Group, East Sussex

www.ribapublishing.com